NATIONAL AUDUBON SOCIETY POCKET GUIDE

A Chanticleer Press Edition

Ann H. Whitman
Editor

Jerry F. Franklin
John Farrand, Jr.
Consultants

Western Region

FAMILIAR TREES OF NORTH AMERICA

Alfred A. Knopf, New York

This is a Borzoi Book.
Published by Alfred A. Knopf, Inc.

 Published in the United States by Alfred A. Knopf, Inc., New York, and simultaneously in Canada by Random House of Canada Limited, Toronto. Distributed by Random House, Inc., New York.

Prepared and produced by Chanticleer Press, Inc., New York.
Color reproductions by Nievergelt Repro AG, Zurich, Switzerland.
Typeset by Dix Type Inc., Syracuse, New York.
Printed and bound by Toppan Printing Co., Ltd., Tokyo, Japan.

Published October 1986
Fifth Printing, February 1995

Library of Congress Catalog Card Number: 86-045584
ISBN: 0-394-74852-2

Contents

How to Use This Guide

Trees are our largest and most conspicuous plants, a dominant part of nearly every landscape. Learning to identify trees will not only add to your appreciation of the American countryside, but will also serve as a good introduction to other aspects of nature, since trees usually have a great influence on the kinds of birds, wildflowers, mammals, and other wildlife in an area.

Coverage

This new guide covers 80 of the most common and frequently encountered tree species in the West. Our range is bounded by the Arctic tree line across northern Canada, the Pacific Ocean on the west, Mexico to the south, and the eastern foothills of the Rockies to the east. The companion volume to eastern trees covers species east of this boundary.

Organization

This easy-to-use guide is divided into three parts: introductory essays, illustrated accounts of the trees, and appendices.

Introduction

As a basic introduction, the essay "Identifying Trees" suggests questions to ask yourself when you see an unfamiliar tree. "Key Features of a Tree" describes and illustrates a tree's characteristic elements—leaf shapes, fruits and cones, and representative winter silhouettes.

An awareness of these features is essential to tree identification.

The Trees

This section contains 80 color plates arranged visually by the overall shape of a tree and its leaf. The first group includes coniferous trees and the second, broadleaf hardwoods. Facing each color plate is a description of the tree's most important features, such as its flowers and fruit, habitat, geographic range, and elevation. The introductory paragraph for each species discusses the uses of the wood, folklore, and other subjects. A black-and-white silhouette of the tree supplements the photograph. For evergreens, the silhouette shows the tree's year-round appearance; for trees that shed their leaves, winter silhouettes show the basic tree shape. A close-up of the bark—often an important identifying feature—is also included with each account.

Appendices

Featured here is an essay on the 20 common families of trees in our area. Knowing family traits helps to recognize many related species.

Whether you live in the country, surrounded by peaceful woods, or in the city, this guide is certain to bring you pleasure and a deeper understanding of nature.

Identifying Trees

The most important step in identifying trees is learning what features to look for. This "questionnaire" is designed to help teach you the points to consider when you look at a tree; in a step-by-step fashion, you will go from the general to the specific and soon narrow the identification to a few clear choices. Remember to take into account the habitat and range of a tree as well as its physical characteristics when you read the text descriptions in the book.

Conifers

If the tree is a conifer, are the leaves shaped like needles (pines, firs, spruces, larches, and others), scales (cypresses), or awls (junipers)? If needlelike, are they borne in clusters (pines, larches) or singly? If in clusters, how many are in a bunch? If the needles are borne singly (hemlocks, spruces, firs, Douglas-firs), are they sharply pointed or rounded on the end?

Once you have examined the needles, look at the cones. Are they upright or pendent? Small or large? Where on the tree are they growing?

Broadleaf Hardwoods

If the tree is a hardwood, are the leaves simple or compound? If they are simple, are they opposite (maples), or alternate (oaks, elms, poplars)? Are the leaves lobed? Are the lobes palmate (maples,

Sweetgum?) Or pinnate (oaks)? If the leaves are not lobed, are they toothed (elms, poplars, birches)? Are the leaves evergreen (live oaks, rhododendrons, magnolias)? Do they have spiny margins (hollies)? Is the bark deeply furrowed (Yellow-poplar, many oaks), smooth and green or white (poplars), or papery and peeling (birches)? Are there large, showy flowers (magnolias, rhododendrons, Yellow-poplar)? What kind of fruit is present? A winged key (maples)? A dense ball of seeds (Sweetgum, sycamores)? A conelike cluster of seeds (birches)? An acorn (oaks)? A stone fruit (cherries, hawthorns)?
If the leaves are compound, are they pinnately compound (ashes, sumacs, locusts, hickories) or palmately compound (buckeyes)? If the leaves are pinnately compound, are they opposite (ashes) or alternate (hickories, locusts)? Do the stems have thorns (locusts)? What kind of fruit is present? Is the plant shrubby (sumacs)? Is the tree growing in a swamp and does it have a swollen or fluted base (baldcypresses, tupelos)?

Key Features of a Tree

In learning to identify trees, it is important to become acquainted with the major elements of each species—its leaf shape, fruit or cones, and silhouette. The pages that follow present examples of the diversity of these significant characteristics.

Leaf Shapes

Recognizing leaf shapes at a glance is a quick way to become comfortable with identifying trees. In some cases, family members have similarly shaped leaves; for example, if you can recognize the classic "maple leaf," you will be able to identify not only all the maples in this book but also a variety of similar relatives. On the pages that follow, 10 of the most common leaf shapes are illustrated for you.

Fruits and Cones

Another good way to recognize a tree is to learn what its fruit looks like. This is especially useful for some deciduous trees in late fall or early winter, when the fruit may persist after the leaves have fallen. If leaves are still present, you can use the fruit to confirm your identification. In some large groups, such as the oaks, identifying the fruit may be the simplest and surest way to distinguish similar species.

The fruit of a tree is a developed, fertilized ovary that contains the seed or seeds. Because fruits develop from flowers, they reflect, in their arrangement, the arrangement of the flowers of that species. Thus trees that bear solitary flowers, such as magnolias, also bear solitary fruits; those species that bear clusters of flowers, such as hollies, also have clusters of fruits.

Some fruits, like cherries or elm keys, are one-seeded; others, such as mulberries or the fruits of magnolias, contain hundreds of seeds. Conifers bear their seeds within cones, which differ from the fruits of broadleaf trees in several important scientific ways. Conifer cones typically contain hundreds of seeds.

A tree may bear nuts, berries, drupes, capsules, or some other kind of fruit. On pages 14 and 15, you will find illustrated 10 different fruit types.

Silhouettes

Like learning the shape of a leaf or the kind of fruit a tree bears, recognizing its overall shape, or silhouette, is often a means to identification, for many trees have a characteristic shape. On pages 16 and 17, you will find illustrations of five typical silhouettes—winter shapes for deciduous trees and year-round silhouettes for evergreens—that will introduce you to the principles of recognizing a tree by its shape.

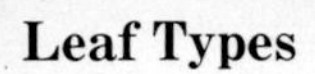

Leaf Types

Scales

Needles in Bundle

Elliptical

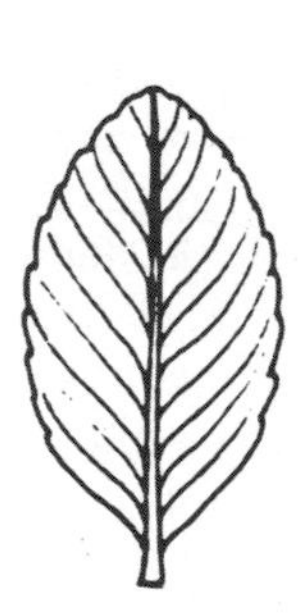

Lobed

Lance-Shaped

Oblong

Oval

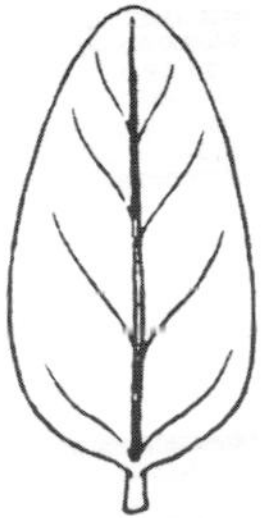

Palmate

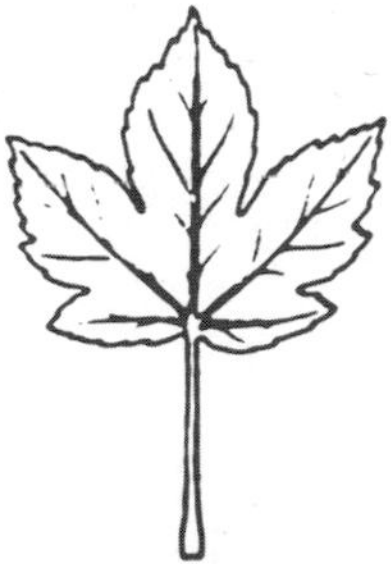

Palmately Compound

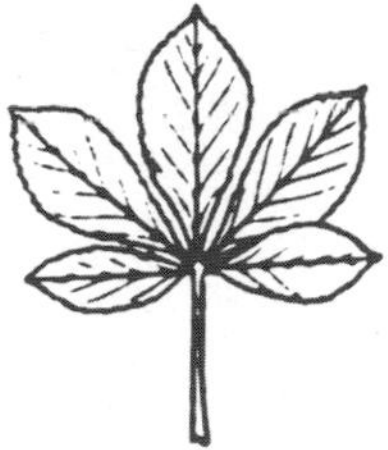

Pinnately Compound

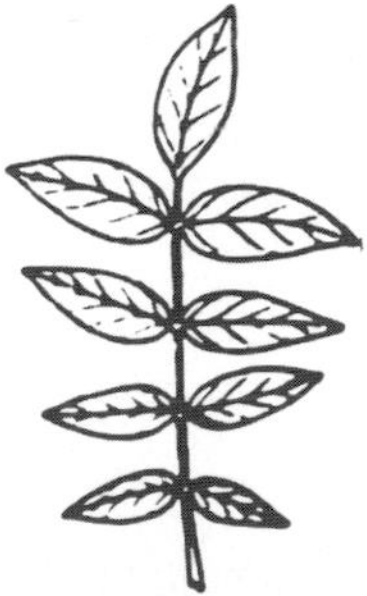

Types of Fruits and Cones

Cone
pines, firs, hemlocks, spruces, larches

Nut
chestnuts, hickories, beeches, buckeyes

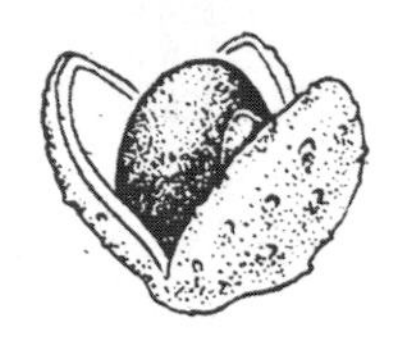

Pod
locusts, acacias

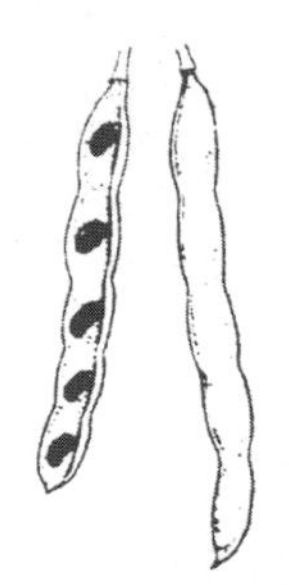

Double key (samara)
maples

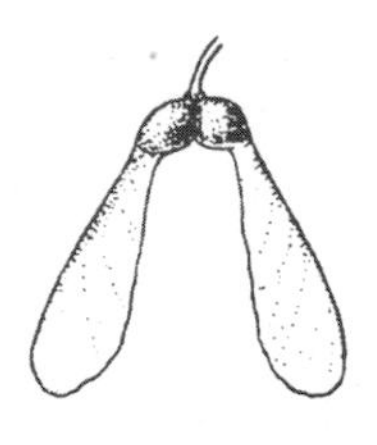

Acorn
oaks, Tanoak

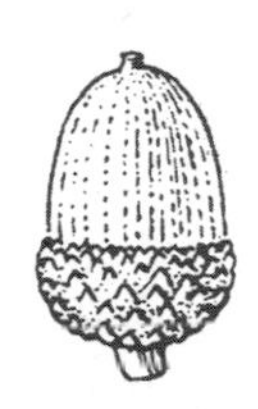

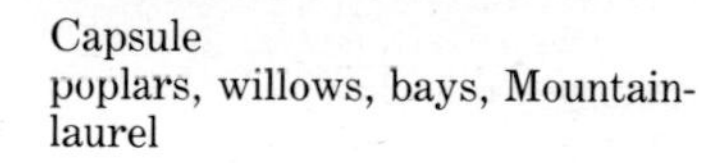

Capsule
poplars, willows, bays, Mountain-laurel

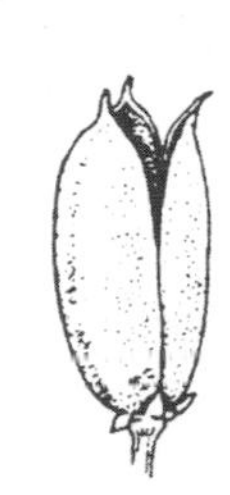

Multiple of nutlets
Sycamore, Sweetgum, birches, Yellow-poplar

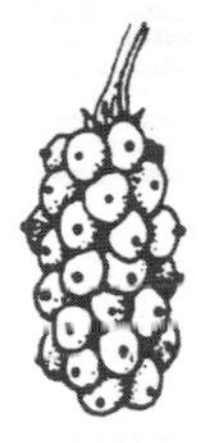

Drupe
cherries, plums, hawthorns, peaches

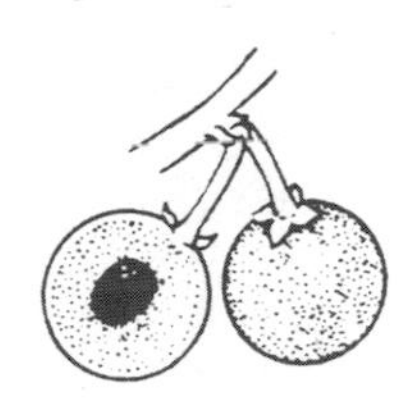

Berry
hollies, dogwoods, elders

Pome
apples, pears

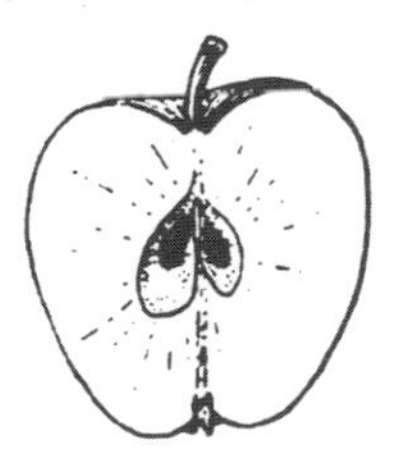

Tree Silhouettes

Narrow, pointed crown of spreading branches

Straight, narrow crown of horizontal branches

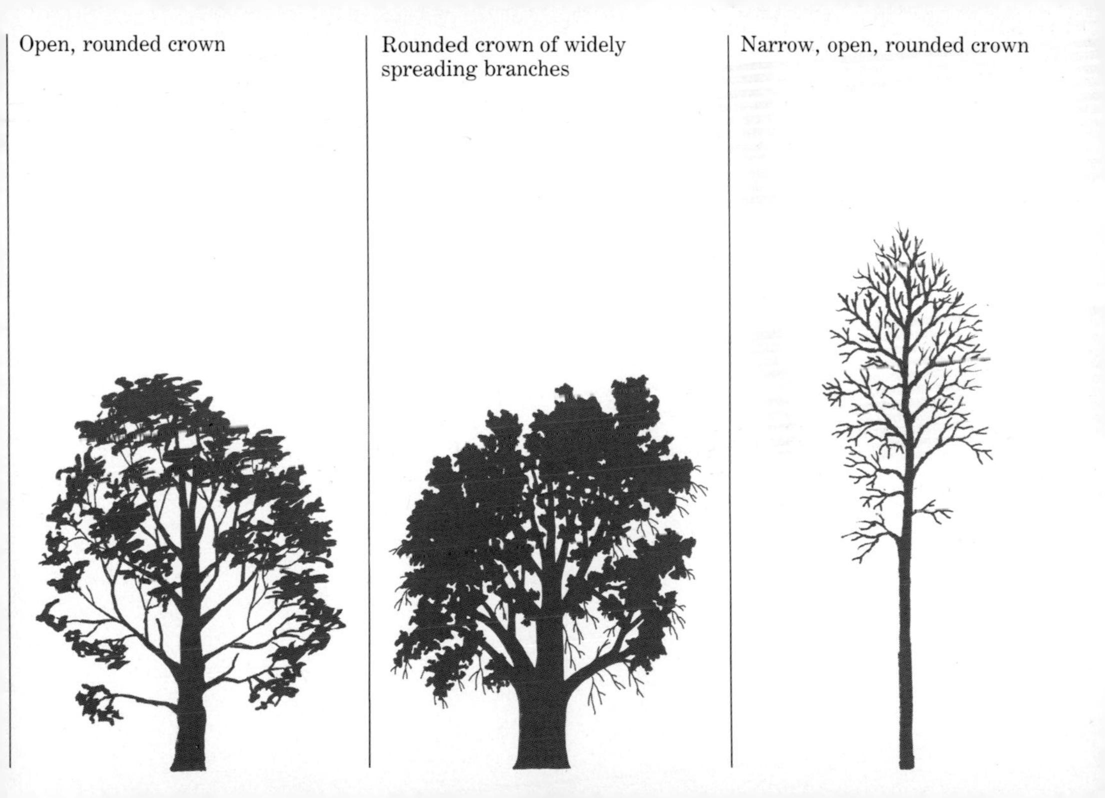
Open, rounded crown
Rounded crown of widely spreading branches
Narrow, open, rounded crown

THE TREES

Monterey Cypress *Cupressus macrocarpa*

Growing along seaside cliffs in California, this striking cypress is easy to recognize; it looks almost as though its image has been distorted through a fun-house mirror. Young Monterey Cypresses, before they are exposed to the buffeting and sculpturing forces of sea winds, have a very different, symmetrical shape. Although widely planted as an ornamental, this species has a very restricted natural range; it is native to only two groves in California.

Identification — Height: 60–80′; diameter: 2′; occasionally larger. Medium-size evergreen with large, flat-topped, irregular crown. Leaves bright green, scalelike, 1⁄16″ long or longer; opposite in 4 rows; blunt. Cones brown, rounded, slightly longer than wide; 1–1⅜″ long; 8–12 cone-scales with stout point; remaining closed and attached.

Habitat — Coastal areas and headlands exposed to spume and sea winds.

Range — Monterey County, California; widely planted as an ornamental outside native range.

Utah Juniper *Juniperus osteosperma*

The Utah Juniper can often be recognized by the presence of a white-berried parasite—mistletoe—which is adapted to living off the bounty of this species. Visitors to the Grand Canyon will be familiar with the Utah Juniper, which grows conspicuously along the South Rim. Like other junipers, this species produces berrylike fruit (technically a cone) that is consumed in large quantities by wildlife.

Identification — Height: 15–40′; diameter: 1–3′. Small to medium-size evergreen with short trunk and open, rounded or conical crown. Leaves yellow-green, awl-like; 1/16″ long; usually growing opposite, in 4 rows along twigs. Cone berrylike, bluish, with white bloom; 1/4–5/8″ in diameter; becoming brownish and dry with maturity.

Habitat — Plateaus, hills, canyons, and dry plains; usually in rocky soils; often with pinyons, but also forming pure stands.

Range — Nevada to Wyoming and south to S. California and New Mexico; local in S. Montana; 3000–8000′.

Rocky Mountain Juniper *Juniperus scopulorum*

Like other junipers, this species is closely related to the cypresses. The two genera can usually be distinguished by looking at the cones: in *Juniperus*, these look more like berries than they do cones. Rocky Mountain Juniper is a popular ornamental, and several horticultural varieties have been developed. Like its close relative the Eastern Redcedar (*J. virginiana*), Rocky Mountain Juniper is an important food source for wildlife.

Identification: Height: 20–50′; diameter: 1½′. Small to medium-size evergreen with narrow, pointed crown and aromatic leaves. Leaves gray-green, awl-like, 1⁄16″ long; opposite in 4 rows along twigs; pointed. Cones berrylike; blue, with a whitish bloom; resinous; maturing in second year. Male (pollen) cones borne on separate trees.

Habitat: Mountainous areas with open woodlands, often on limestone soils or lava extrusions; forming foothill woodlands with pinyons.

Range: Central British Columbia south through mountains to Trans-Pecos Texas; 5000–9000′ (almost to sea level in north).

Western Redcedar *Thuja plicata*

Because its wood is durable and resistant to decay, Western Redcedar is a very important species in the manufacture of shingles, paneling, fence posts, and other outdoor objects; it is also used in boat building. The fabulous totem poles carved by Native Americans of the Pacific Northwest are made from trunks of this tree, also known as Canoe-cedar and Giant Arborvitae.

Identification: Height: 100–175′; diameter: 2–8′; sometimes larger. Large evergreen with massive trunk and narrow, conical crown; branches droop at ends. Leaves shiny dark green, usually with some white marks below; scalelike; 1/16–1/8″ long, opposite in 4 rows, with short point at tip; side pairs of leaves keeled. Cones brown, elliptical, ½″ long, growing in clusters upright from small stalk; with 10–12 paired, leathery cone-scales.

Habitat: Slightly acid, moist soils, often with Western Hemlock in large forests; also with other conifers.

Range: SE. Alaska along coast to NW. California; also in Rockies from SE. British Columbia to W. Montana; to 3000′ in north; to 7000′ in south.

Alaska-cedar *Chamaecyparis nootkatensis*

This species, which is also known as Alaska Yellow-cedar and Nootka Cypress, has strong wood that is used today to make furniture and boats. Native Americans of the Pacific Northwest made ceremonial masks from the wood of Alaska-cedar. The crushed leaves of this popular ornamental have a pungent odor, and the lumber is fragrant.

Identification — Height: 50–100′; diameter: 1–4′. Medium-size to tall evergreen with horizontal or drooping branches forming a narrow crown. Leaves scalelike, bright yellow-green; ⅛″ long; opposite, and growing in 2 rows along twigs. Cones reddish-brown, rounded, ½″ in diameter; 4–6 rounded cone-scales with a long point at tip; maturing in second year.

Habitat — Mountain areas with wet soils; mainly with other conifers but sometimes in pure stands.

Range — SE. Alaska along coast to mountains of W. Oregon and NW. California; local inland. From sea level in north; 2000–7000′ farther south.

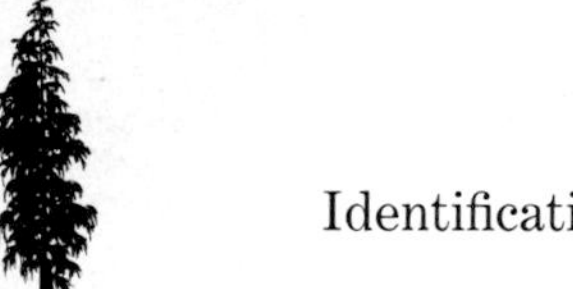

Port-Orford-cedar *Chamaecyparis lawsoniana*

Port-Orford-cedar produces fragrant wood that is used to make boats, flooring, lacquerware and toys; in Japan, the wood is also commonly used in the construction of shrines and temples. Also known as Oregon-cedar, Lawson Cypress, and Lawson's False-cypress, this species is native only to a tiny area along the fogbound Pacific Coast, but there are as many as 100 horticultural varieties, and so the tree is more famous than its small native range would indicate.

Identification — Height: 70–200′; diameter: 2½–4′. Tall evergreen with narrow, spirelike crown and drooping branches. Leaves dull green above, whitish below; scalelike, 1/16″ long; opposite, in 4 rows along twig. Cones reddish brown, often with whitish bloom; rounded, 3/8″ in diameter; with 8–10 blunt cone-scales; maturing in first season.

Habitat — Clayey or sandy loams and rocky ridges; often with other conifers but sometimes in pure stands.

Range — Narrow coastal area in SW. Oregon and NW. California; local near Mt. Shasta; to 5000′; cultivated forms widely planted.

Incense-cedar *Libocedrus decurrens*

A popular ornamental, Incense-cedar is also an important producer of timber; the wood, which is used to make fragrant cedar chests and closets, is additionally very well suited to the manufacture of pencils because it does not splinter. Mature specimens, however, sometimes fall prey to a fungus infection that rots the wood. This species is the only representative of its genus in North America.

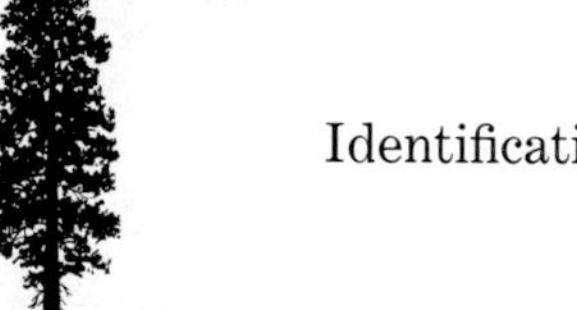

Identification: Height: 60–150′; diameter: 3–5′. Large evergreen with tapering, angular trunk; crown narrow and conical in young trees, becoming irregular with age. Leaves shiny green, scalelike; ⅛–¼″ long, growing opposite in 4 rows along twig; side pair of leaves keeled and overlapping with other pairs along twig. Cones reddish-brown, oblong, ¾–1″ long; hanging from leafy stalk; with 6 hard, flat, pointed cone-scales.

Habitat: Mountainous areas; usually with other conifers, but occasionally in pure stands.

Range: W. Oregon to S. California and extreme W. Nevada; 1200–7000′.

Giant Sequoia *Sequoiadendron giganteum*

The Giant Sequoia, one of the tallest trees in the world, is also one of the longest-lived; mature Sequoias are typically more than 1500 years old. Because the species is rare, it is carefully protected today in state and federal parks and forests. Fortunately, this species is resistant to fire, insects, and disease—a factor that probably contributes to its longevity.

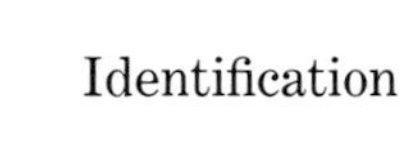

Identification Height: 150–250′; diameter: 20′; sometimes much larger. Massive evergreen with reddish-brown, tapering trunk and narrow, conical crown of horizontal branches; very old trees have irregular, open crown. Leaves blue-green with 2 whitish lines; scalelike and overlapping, ⅛–¼″ long; oval or lance-shaped, with sharp-pointed tip. Cones reddish brown, elliptical; 1¾–2¾″ long; hanging from twigs; with flat, short-pointed cone-scales; maturing in second year and remaining attached.

Habitat Moist mountain slopes with rocky soils; in groves with other conifers.

Range Western slope of Sierra Nevada in central California; 4500–7500′, rarely higher or lower.

Pacific Yew *Taxus brevifolia*

Like other yews, the Pacific Yew is poisonous; eating the foliage or seeds can result in death. Despite the small size of the tree, the wood is useful in making canoe paddles, poles, and archery bows; indeed, before the advent of gunpowder, yew wood was so important for making weapons that the trees were protected by law in much of Europe.

Identification — Height: 50′; diameter: 2′. Small evergreen tree (sometimes a shrub) with angular, twisted trunk and broad crown of slender branches. Needles dark yellowish green above, light green with 2 whitish bands below; flat, short-pointed, flexible, and soft. Male cones and seeds on different trees; seeds berrylike, scarlet, 3⁄8″ in diameter; male cones pale yellow, with short stalk; 1⁄8″ in diameter.

Habitat — Stream banks and canyons with moist soils; with other conifers.

Range — Extreme SE. Alaska along coast to central California; also in Rockies from SE. British Columbia to N. Montana and central Idaho; at sea level north; to 7000′ in south.

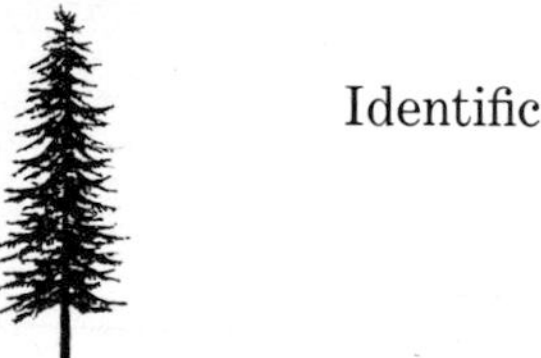

Redwood *Sequoia sempervirens*

The Redwood is the tallest tree in the world. The species reaches maturity after 400 to 500 years; the oldest known Redwood lived 2200 years. Before the arrival of European settlers, this majestic tree occupied vast areas, but the newcomers were quick to decimate these virgin forests. The genus was named in honor of Sequoyah, inventor of the Cherokee alphabet.

Identification: Height: 200–325′; diameter: 10–15′; sometimes much larger. Huge evergreen with massive reddish-brown trunk and short, irregular crown of horizontal, drooping branches. Leaves of 2 kinds: needles dark green above, whitish-green below, ⅜–¾″ long; growing in 2 rows; flat, somewhat curved, with pointed tip; leaves on leader (top branch) tiny, scalelike, with keel; spreading around twig. Cones red-brown, elliptical, pendent; ½–1⅛″ long; cone-scales flat with short point; maturing in first season.

Habitat: Cool, humid areas with alluvial soils, often on terraces and flats; in pure stands or with other conifers.

Range: Extreme SW. Oregon along coast to central California; sea level to 3000′.

Western Hemlock *Tsuga heterophylla*

Western Hemlock is the tallest member of its genus and a very important source of timber and pulpwood. This species is one of the most common trees in the great forests of the Pacific Northwest, where it may occur alone in huge groves or with Sitka Spruce and other conifers. It is not related to Poison Hemlock (*Conium maculatum*), the Old World herb that was used to poison Socrates.

Identification — Height: 100–150′; diameter: 3–4′. Tall, slender evergreen with narrow, conical crown and drooping leader (topmost branch). Needles dark green above, with 2 whitish bands below; ¼–¾″ long, growing in 2 rows along twigs; flat, short-stalked, and flexible. Cones brown, elliptical; ¾–1″ long; stalkless, hanging down from twigs; with rounded, elliptical cone-scales.

Habitat — Flat areas and lower slopes with moist, acid soils; often in pure stands.

Range — S. Alaska along coast to NW. California and in Rockies from SE. British Columbia to N. Idaho and NW. Montana; to 2000′ along coast, and to 6000′ in mountains.

Grand Fir *Abies grandis*

Despite its great size—Grand Fir has been known to reach a height of nearly 300 feet, with a diameter of six feet—this tree produces weak, coarse-grained wood that is not particularly useful. Like other firs, Grand Fir has winged seeds that are a valuable staple in the diets of many songbirds and small mammals.

Identification Height: 100–200′; diameter: 1½–3½′. Tall evergreen with pointed, narrow crown of slightly drooping branches. Needles dark green and shiny above, whitish below; 1¼–2″ long, growing in 2 nearly perpendicular rows along twigs; flat, flexible. Cones green or greenish brown; cylindrical, 2–4″ long, growing upright on topmost twigs; cone-scales hairy with short bracts.

Habitat Cool, humid areas, especially valleys and mountainsides; in coniferous forests.

Range S. British Columbia along Pacific Coast to California; also in Rockies to central Idaho; to 1500′ in coastal areas, 6000′ inland.

White Fir *Abies concolor*

This large fir has two geographical varieties. Rocky Mountain White Fir (var. *concolor*) grows in warm, dry areas; California White Fir (var. *lowiana*) is found along the Pacific Coast. White Fir, like other members of its genus, produces a fragrant resin, which rises in small blisters on the trunk.

Identification Height: 70–160′; diameter: 1½–4′. Tall evergreen with short, horizontal branches forming pointed, narrow crown. Needles light bluish green with white lines above and below; 1½–2½″ long, in 2 nearly perpendicular rows; flat, flexible. Cones greenish, yellow, or purplish; 3–5″ long and cylindrical, growing upright from topmost twigs; cone-scales with fine hairs and short, concealed bracts.

Habitat Mountain areas with rocky, moist soils; in pure stands and with other firs.

Range SW. Oregon and extreme SE. Idaho south as far as S. California and S. New Mexico; 5500–11,000′ in south; to 2000′ in north.

Engelmann Spruce *Picea engelmannii*

At high elevations, Engelmann Spruce and Subalpine Fir grow together in vast, uniform stands; these spruce-fir forests are among the most important forest types of the West. The wood of Engelmann Spruce is used to make violins and piano sounding boards. The needles, when crushed, give off an unpleasant, skunklike odor.

Identification: Height: 80–100′; diameter: 1½–2½′. Tall evergreen with short branches forming a dense, conical crown. Needles dark green or blue-green with whitish lines; ⅝–1″ long, growing on all sides of twig; 4-sided, with sharp point; flexible, slender. Cones light brown, shiny; cylindrical, 1½–2½″ long, hanging at end of twig; cone-scales flexible, thin, with irregular teeth.

Habitat: Mountain areas with Subalpine Fir; also with other conifers.

Range: Central British Columbia and SW. Alberta in mountains to Arizona and New Mexico; 8000–12,000′ in south; from 2000′ in north.

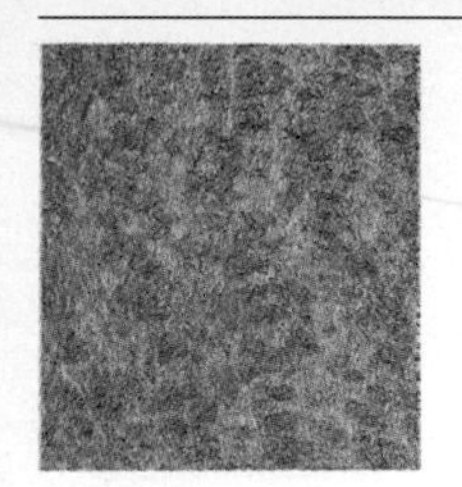

White Spruce *Picea glauca*

Like its relative the Black Spruce, with which it forms vast forests across the Far North, the White Spruce often occurs as a small shrub at treeline. This species is tremendously important in Canada; it is a major source of pulpwood and construction lumber, and is also used to make violins and other musical instruments.

Identification: Height: 40–100′; diameter: 1–2′. Evergreen with conical crown of horizontal branches. Needles blue-green with whitish lines, ½–¾″ long; 4-sided and growing mainly from upper surface of twig; sharp-pointed; crushed needles give off a skunklike odor. Cones light brown and shiny; 1½–2½″ long; cylindrical, with thin, flexible cone-scales; hanging at ends of twigs and falling at maturity.

Habitat: Coniferous forests with a variety of soil types; often in pure stands or with Black Spruce.

Range: From northern limit of trees in Alaska, N. Quebec, and Newfoundland south to British Columbia, the Great Lakes, and Maine; to timberline (2000–5000′) in mountains.

Sitka Spruce *Picea sitchensis*

Sitka Spruce is one of the world's tallest and fastest-growing spruces, often adding three feet to its height each year. It takes its name from the former Russian colonial capital of Alaska. The light, strong lumber is used in making boats and piano sounding boards; in the early days of aviation, it was used to build airplanes.

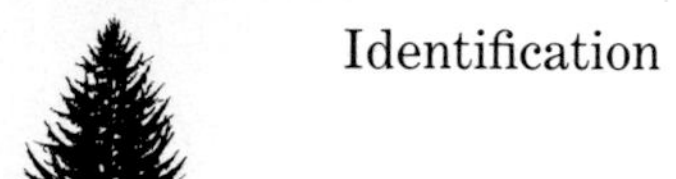

Identification — Height: 160′; diameter: 3–5′; sometimes much larger. Tall evergreen with straight trunk, buttressed base, and broad, conical crown. Needles dark green; ⅝–1″ long; growing from all sides of twig; flattened with sharp point at tip and slight keel along lower surface. Cones light orange-brown, cylindrical; 2–3½″ long; with thin, stiff cone-scales; hanging at ends of twigs, opening and falling at maturity.

Habitat — Humid, foggy areas of the northwest coastal forest, in cool, moist sites; often in pure stands; sometimes with Western Hemlock.

Range — S. Alaska to NW. California; to timberline at 3000′ in Alaska; to 1200′ in California.

Black Spruce *Picea mariana*

Found in the East as well as the West, the Black Spruce often grows with the White Spruce (*P. glauca*). The leaves of the Black Spruce are used to make spruce beer; the wood, although used locally, is of little economic value because of the tree's small size. When heavy snow bends the branches to the ground, they take root, producing a group of young trees around the older one.

Identification — Height: 20–60′; diameter: 4–12″. Small to medium-size evergreen with irregular crown; often a small shrub at timberline. Needles blue-green, with whitish lines; ¼–⅝″ long; 4-sided with a sharp point; growing on all sides of twig. Cones dull gray, egg-shaped; ⅝–1¼″ long; with stiff, rounded cone-scales; curving downward from short stalk; often in clusters near top of tree; remaining attached after maturing.

Habitat — Conifer forests and other areas with wet soils, such as peat bogs and loamy areas; often in pure stands.

Range — Alaska to Labrador at northern limit of trees; south through Canada to British Columbia, Minnesota, and New Jersey; 2000–5000′.

Douglas-fir *Pseudotsuga menziesii*

This magnificent tree produces fine lumber, and it is one of the leading timber trees of the world. Douglas-fir can be distinguished from true firs by its hanging cones, which have long, protruding bracts. There are two geographical varieties, Coastal Douglas-fir (var. *menziesii*) and Rocky Mountain Douglas-fir (var. *glauca*).

Identification — Height: 80–200′; diameter: 2–5′. Large evergreen with slightly drooping branches forming a narrow, pointed crown. Needles dark yellow-green or blue-green; ¾–1¼″ long, growing in 2 rows; flat, flexible, with rounded tip and short, twisted stalk. Cones light brown, narrowly egg-shaped; 2–3½″ long; with thin cone-scales and protruding bracts; hanging from short stalks.

Habitat — Coastal variety on moist, well-drained soils, and often in pure stands; Rocky Mountain variety in mountainous areas with rocky soils, often with other conifers but also in pure stands.

Range — Central British Columbia along coast to central California, and in Rockies to SE. Arizona; from 2000′ in north, and from 8000–9500′ in south; local elsewhere.

Blue Spruce *Picea pungens*

The needles of this high-altitude spruce have a waxy resin; this material enables these trees to withstand hot, droughty summers and cold winters alike. Cultivated varieties have beautiful bluish-white foliage. The Blue Spruce is a favorite Christmas tree; because it is slow-growing, it has also attained some popularity as an ornamental.

Identification — Height: 70–100′; diameter: 1½–3′. Large evergreen with striking bluish foliage and conical crown of horizontal branches. Needles blue-green or bluish (sometimes silvery); ¾–1⅛″ long; 4-sided, with sharp point at tip, growing from all sides of twig; crushed needles have resinous odor. Cones light brown, shiny; 2¼–4″ long; cylindrical; cone-scales thin, long, flexible.

Habitat — Mountain streamsides and bottomlands with hot, sunny summers and cold winters; often in pure stands.

Range — In the Rockies from S. and W. Wyoming and E. Idaho to NE. Arizona and S. New Mexico; 6000–11,000′.

Mountain Hemlock *Tsuga mertensiana*

Also known as Black Hemlock and Alpine Hemlock, this species grows in high mountains, and it often appears as a deformed-looking shrub at timberline. The wood is of little value, but the tree has become popular as an ornamental, particularly in the East. The leaves and seeds are an important food source for wildlife.

Identification Height: 30–100′; diameter: 1–3′. Evergreen of varying size with slender branches forming a conical crown, with drooping leader (topmost branch). Needles blue-green, with whitish lines above and below; growing on all sides of twigs and crowded at ends of side twigs; flat above, with blunt tip and short stalk. Cones purplish to brown, cylindrical, 1–3″ long; stalkless and hanging; with rounded cone-scales.

Habitat With firs and in mixed coniferous forests on moist, coarse or rocky soils.

Range W. Alaska to British Columbia; in mountains to central California, and in Rockies to NE. Oregon and N. Idaho; to 3500′ in north and 5500–11,000′ farther south.

Subalpine Fir *Abies lasiocarpa*

The narrow, graceful spires of this fir are a familiar sight in the Rockies, where their form seems to reflect the magnificence of the surrounding mountain peaks. Subalpine Fir is the most widespread true fir of the West, and it is an important food source for a variety of wildlife; deer, bighorn sheep, and elk browse the bark, while songbirds and small mammals consume the seeds.

Identification: Height: 50–100′; diameter: 1–2½′. Tall evergreen with dense, pointed crown reaching nearly to base of trunk. Needles dark green with whitish lines above and below; flat, 1–1¾″ long, growing in 2 nearly perpendicular rows. Cones dark purple, cylindrical, 2¼–4″ long; growing upright on topmost twigs; cone-scales hairy with short, concealed bracts.

Habitat: Just below timberline (subalpine forest zone); often with Engelmann Spruce, forming spruce-fir forests; also with other conifers.

Range: SE. Alaska to S. New Mexico; from near sea level in northernmost part of range; 8000–12,000′ in south.

California Red Fir *Abies magnifica*

The scientific name of this fir is appropriate; in maturity, the California Red Fir is majestic and beautifully symmetrical. The common name is a reference to the characteristic reddish-brown bark of the species. On high western slopes of the Sierra Nevada, this large, handsome tree can be found in almost pure forests, forming a distinct zone.

Identification: Height: 60–120′; diameter: 1–4′. Large evergreen with narrow branches forming conical crown, rounded at top. Needles bluish green with whitish lines; ¾–1⅜″ long; growing in 2 rows along twig; 4-sided. Cones purplish brown, large (6–8″ long) and cylindrical; growing upright from upper branches; cone-scales finely hairy, pointed, with fine teeth.

Habitat: High-altitude areas with deep winter snowfall and dry summers; often in pure stands.

Range: SW. Oregon in the Cascades to Coast Ranges of California, through Sierra Nevada to central California and W. Nevada; 6000–9000′ in south; to 4500′ in north.

Limber Pine *Pinus flexilis*

This tree is aptly named; its tough, pliant twigs are so flexible that they can sometimes be tied in a knot. In exposed areas, the wind often shapes these trees into stunted, deformed-looking shrubs. Like other pines, it produces a new row of branches every year. The large seeds of this species are an important food for many kinds of wildlife.

Identification: Height: 40–50′; diameter: 2–3′. Medium-size evergreen with stout branches forming rounded crown. Needles green with white lines; 2–3½″ long; slender, in bundles of 5; first-year needles sheathed. Cones yellow-brown, egg-shaped; 3–6″ long, with thick, rounded, and bluntly pointed cone-scales; opening at maturity.

Habitat: High mountain ridges and dry, rocky slopes; often in pure stands.

Range: In the Rockies from SE. British Columbia and SW. Alberta to N. New Mexico and S. California; local in the Dakotas and W. Nebraska; 5000–12,000′.

Whitebark Pine *Pinus albicaulis*

The cones of the Whitebark Pine do not open; instead, they are shed at maturity, and the seeds are released only when the cone decays. Scientists believe that this method of seed dispersal is evidence that the Whitebark is a very ancient and primitive pine. The seeds are a favorite food of the noisy, intrepid Clark's nutcracker, a jaylike bird of western mountains.

Identification: Height: 20–50′; diameter: 1–2′. Small evergreen with short, crooked trunk and irregular crown. Needles dull green with faint white lines; 1½–2¾″ long, in bundles of 5; stout, stiff; crowded at ends of twigs; first-year needles sheathed. Cones purple to brown, egg-shaped or rounded; 1½–3¼″ long; with very thick, pointed cone-scales; shedding at maturity.

Habitat: High-elevation and northern slopes and ridges with dry, rocky soils; sometimes in pure stands.

Range: Central British Columbia and SW. Alberta to central California and W. Wyoming; 4500–7000′ in north; 8000–12,000′ in south.

Pinyon *Pinus edulis*

The "pine nuts" of this species are used for a variety of foods. Eaten raw, they are delicious in salads; they may also be baked into cookies or ground and used with fresh basil to make pesto, a kind of sauce for spaghetti. Unlike most nut-producing trees in North America, the Pinyon is not cultivated; its tasty bounty must be gathered in the wild.

Identification: Height: 15–35′; diameter: 1–2′. Small evergreen with short trunk and rounded, compact crown. Needles light green, stout; ¾–1½″ long, usually in bundles of 2 (but sometimes 1 or 3). Cones yellow-brown, resinous; egg-shaped, 1½–2″ long, with thick, blunt cone-scales; opening and shedding at maturity.

Habitat: Woodlands and open areas, and dry, rocky foothills; often with junipers.

Range: Southern Rockies of Utah, Colorado, New Mexico, and Arizona; local elsewhere in the Southwest; mostly from 5000–7000′.

Monterey Pine *Pinus radiata*

The fire-adapted Monterey Pine is native to California, where it is comparatively rare and local; but it has been successfully introduced to the Southern Hemisphere and is grown commercially there for its valuable timber. A popular ornamental in the Pacific states, this species has also been introduced to some areas of southern Europe.

Identification: Height: 50–100′; diameter: 1–3′. Medium-size to tall evergreen with irregular, open crown and straight trunk. Needles shiny green, slender; 4–6″ long, in bundles of 3. Cones shiny brown, egg-shaped or conical, appearing 1-sided at base; 3–6″ long, in clusters or whorls; cone-scales thick, raised, rounded. Cones remain closed on tree for many years, opening in response to heat or fire.

Habitat: Slopes with coarse, somewhat sandy soils; often in pure stands; also occurs with Monterey Cypress and Coast Live Oak.

Range: Coastal central California, near Monterey; to 1000′; introduced elsewhere.

Digger Pine *Pinus sabiniana*

This pine can be distinguished at once by its foot-long, drooping, gray-green leaves. It also has a distinctive, forking trunk; because of its shape, the wood that the tree produces is not commercially valuable, although it is widely used as fuel. The name "Digger" comes from a 19th-century term applied by settlers to all the Native American tribes of California, who formerly used the seeds for food.

Identification — Height: 40–70′; sometimes much larger; diameter: 2–4′. Medium-size to tall evergreen with forked trunk and thin, irregular crown. Needles dull gray-green with white lines; 8–12″ long; in bundles of 3. Cones brown, egg-shaped, 6–10″ long; heavy, on long, bent stalks; cone-scales long, thick, with sharp keel; 4-sided, with a large, stout spine; opening late and persisting for many years.

Habitat — Foothills and low mountains on dry slopes and ridges; with oaks and other conifers.

Range — Coast Ranges and Sierra Nevada of California; 1000–3000′ (rarely higher or lower).

Knobcone Pine *Pinus attenuata*

As the common name implies, Knobcone Pine bears knobby, irregular cones that remain on the tree for decades; these cones are an easy field mark to recognize. As the trunk expands, the bark sometimes "swallows" the cones, giving the tree a very odd appearance indeed. This species is fire-adapted; the cones open only when exposed to heat, and the seeds take root and grow in burned-over areas.

Identification: Height: 30–80′; diameter: 1–2½′. Medium-size to tall evergreen with narrow, pointed crown, becoming irregular on older trees. Needles yellow-green, 3–7″ long; in bundles of 3; slender, stiff. Cones yellow-brown, shiny, irregularly egg-shaped; 3¼–6″ long, in clusters or whorls; cone-scales raised, keeled, with stout spine at tip; cones remain closed for many years.

Habitat: Mountain areas with thin, rocky soils; often in pure stands.

Range: SW. Oregon to S. California; 1000–2000′ in north; 1500–4000′ in south, sometimes higher.

Sugar Pine *Pinus lambertiana*

The beautiful Sugar Pine is an important producer of lumber. As the West was settled, Sugar Pine was used to build houses; during the gold rush in California, the wood was put to the test in the construction of mine shafts. The cones of the Sugar Pine are longer than those of any other species; the largest ones on record are about 20 inches long.

Identification — Height: 100–160′; diameter: 3–6′; sometimes much larger. Large, tall evergreen with straight trunk unbranched for much of its length; crown open, conical, becoming flat with age. Needles blue-green with white lines; 2¾–4″ long; in bundles of 5; twisted, slender, with sharp point. Cones light brown, shiny; cylindrical, 11–18″ long (sometimes longer); hanging from long stalks on upper branches; cone-scales rounded, with blunt point.

Habitat — Mountain areas with other conifers; on many kinds of soils.

Range — W. Oregon through Sierra Nevada to S. California; 1100–5400′ in north, 2000–7800′ in Sierra Nevada, and 4000–10,500′ in south.

Lodgepole Pine *Pinus contorta*

This species has three distinct geographical varieties. The Shore Pine (var. *contorta*) is a small tree with a spreading crown; unlike most pines, it grows in wet soils. The Sierra Lodgepole Pine (var. *murrayana*) is tall and narrow. Rocky Mountain Lodgepole Pine (var. *latifolia*) is fire-adapted; its cones only open when heated by a forest fire; the tree then pioneers on the burned site.

Identification: Height: 20–80′; diameter: 1–3′. Small to large evergreen, either with tall, narrow crown or short, spreading crown. Needles yellow-green or darker; 1¼–2¾″ long, in bundles of 2; stout, sometimes twisted. Cones yellow-brown, egg-shaped, ¾–2″ long; stalkless, with raised, rounded cone-scales, keeled and with bristle at tip.

Habitat: High altitudes with well-drained soils; Shore Pine in wet, boggy areas and dry, sandy places.

Range: SE. Alaska south along coast to N. California; in Sierra Nevada to S. California, and in Rockies to Colorado; Shore Pine from sea level to 2000′; inland varieties 1500–3000′ in north, and 7000–11,500′ in south.

Jeffrey Pine *Pinus jeffreyi*

Jeffrey Pine can be distinguished by its aromatic bark and twigs, which give off a fragrance like lemons or vanilla. This species was discovered in the 19th century by John Jeffrey, an eminent Scottish botanical explorer, who introduced it to Scotland. The long, pale bluish-green leaves of this species have made it a favorite ornamental.

Identification: Height: 80–130′; diameter: 2–4′; sometimes much larger. Tall evergreen with straight trunk and open, conical crown. Needles blue- or gray-green with white lines; 5–10″ long, in bundles of 3; stout and stiff. Cones light reddish brown; conical or egg-shaped, 5–10″ long; cone-scales raised, keeled, with long prickle; cones opening and shedding at maturity.

Habitat: Dry mountain slopes, especially over lava flows and granite; largest specimens on deep, well-drained soils; often in pure stands.

Range: SW. Oregon through Sierra Nevada to S. central California; usually at 6000–9000′.

Western White Pine *Pinus monticola*

This species, also called Mountain White Pine, is one of the largest members of its genus. It produces high-quality wood, which is used to make matches and wood paneling for the interiors of houses. Like other five-needled pines, the Western White Pine is susceptible to a disease called white pine blister rust, which is caused by an introduced fungus.

Identification: Height: 100′; diameter: 3′; sometimes much larger. Large evergreen with straight trunk and horizontal branches forming a narrow, conical crown. Needles blue-green with whitish lines on inner surface; 2–4″ long, in bundles of 5; first-year needles are sheathed. Cones yellow-brown, cylindrical, 5–9″ long and somewhat slender; most on long stalks; cone-scales thin, rounded, with point at tip; opening and shedding at maturity.

Habitat: Mountainous areas with moist soils; often in mixed forests, sometimes in pure stands.

Range: Rockies from British Columbia to Montana; also in Sierra Nevada to central California; to 3500′ in north; 6000–9800′ in south.

Western Larch *Larix occidentalis*

This species is a close relative of the Tamarack, and is often called Western Tamarack. The Western Larch is a magnificent tree, reaching a height of 150 feet or more; very old trees have bright cinnamon-red bark. The wood is strong and durable, and is used for a great variety of construction purposes. The buds and leaves of this species are an important source of food for grouse.

Identification: Height: 80–150′; diameter: 1½–3′; sometimes much larger. Tall deciduous conifer with narrow conical crown and horizontal branches. Needles light green; 1–1½″ long and very slender; crowded in clusters on spur twigs, or scattered along leader twigs; sharp-pointed, 3-sided; turning yellow in fall. Cones brown; 1–1½″ long; elliptical, growing upright on short stalks; with hairy, irregular cone-scales and longer bracts.

Habitat: Porous, sandy or loamy soils in mountains and valleys; with other conifers.

Range: SE. British Columbia to N. Oregon and NW. Montana; 2000–5500′ in north, to 7000′ in south.

Ponderosa Pine *Pinus ponderosa*

This is the most common and widespread pine in North America; there are several different geographical varieties. Pacific Ponderosa Pine (var. *ponderosa)* is the typical variety, with three needles per bundle; Interior Ponderosa Pine (var. *scopulorum*) has short needles in bundles of two; and Arizona Ponderosa Pine (var. *arizonica*) has five needles per bundle.

Identification — Height: 60–130′; diameter: 2½–4′; sometimes larger. Tall evergreen with spreading branches forming broad, conical crown. Needles dark green; usually 4–8″ long, and usually in bundles of 2 or 3; stout, stiff. Cones light reddish brown; conical or egg-shaped; 2–6″ long, with raised, keeled cone-scales with prickle at tip; opening and shedding at maturity.

Habitat — Mountain areas, forming pure stands and large forests; also in mixed forests with other conifers.

Range — W. British Columbia to SW. North Dakota, south to SW. Texas and S. California; from sea level in north, to 9000′ in south.

Tamarack *Larix laricina*

The Tamarack, also known as Hackmatack and Eastern Larch, looks like an evergreen because it has needles and produces cones. But, like other larches, it sheds its leaves in the fall—an unusual trait among cone-bearing trees. The durable wood of the Tamarack is used to make gates, houses, fence posts, and hulls for fishing boats.

Identification Height: 40–80′; diameter: 1–2′. Medium-size conifer with open, conical crown and thin, straight trunk. Needles light blue-green, soft, slender; ¾–1″ long; in clusters or scattered along twig on leaders; turning yellow in fall. Cones bright red-brown, turning darker brown, with rounded cone-scales; ½–¾″ long; growing upright and falling in second year.

Habitat Bogs and swamps with peaty soils; in uplands with loamy soils; often in pure stands.

Range Central Alaska and N. British Columbia to Labrador, south to Minnesota and N. New Jersey; local farther south. Near sea level in north; to 4000′ in southern part of range.

Black Hawthorn *Crataegus douglasii*

A popular ornamental, the Black Hawthorn bears delicate pale pink or whitish flowers in the spring and shiny purplish-black fruit in late summer. Also called Douglas Hawthorn, it was discovered by David Douglas, a famous Scottish naturalist. It has a more northerly range than any other member of its widespread genus.

Identification: Height: 30′; diameter: 1′. Small tree or shrub with compact crown of short, stout, erect branches. Leaves shiny bright green above, paler below; oval, broader near tip and narrower at base, 1–3″ long and ⅝–2″ wide; with sharp sawteeth, and often with irregular lobes beyond middle. Flowers white, ½″ wide, with 5 petals; stamens pink; in long-stalked clusters 2″ wide; appearing with leaves in spring. Fruit berrylike, ½″ in diameter; dark shiny purple or black; in long-stalked clusters, maturing in late summer.

Habitat: Streamsides and valleys with moist soils; with conifers or sagebrush.

Range: British Columbia along coast to central California, south through Idaho to Colorado and New Mexico; to 6000′.

Western Serviceberry *Amelanchier alnifolia*

Appearing as a shrub or a small tree, the Western Serviceberry is common and widespread, growing in the understory of all the major forest types of the West. A member of the rose family, this species—also called Saskatoon and Western Shadbush—produces abundant, delicious summer fruits that are relished by wildlife and people alike.

Identification Height: 30′; diameter: 8″. Small tree or shrub with several trunks and a narrow, rounded crown. Leaves dark green above, paler below; hairy when young; broadly oval to round, ¾–2″ long and almost as wide; with coarse, jagged sawteeth beyond middle. Flowers white, ¾–1¼″ wide, with 5 petals and bright yellow-green stamens; appearing in clusters in spring with leaves. Fruit like a tiny purple or blackish apple; in long-stalked clusters, maturing in early summer.

Habitat Moist soils in most kinds of forests, and in clearings.

Range Central Alaska and Manitoba south to NW. Iowa, Colorado, and N. California; local elsewhere in East; to 6000′.

Coast Live Oak *Quercus agrifolia*

Early explorers to the California coast called this species the "Holly-leaved Oak" because of its shiny, sharp-toothed leaves. It has a very broad crown, sometimes 100 feet in diameter. Common along the coast south of San Francisco, it forms large groves, often with other oaks. Also called California Live Oak and Encina.

Identification: Height: 30–80′; diameter: 1–3′; sometimes larger. Medium-size, massive evergreen with short trunk and crooked, stout branches forming a large crown; sometimes a shrub. Leaves shiny dark green above, paler below; oblong and rounded at both ends, ¾–2½″ long and ½–1½″ wide; leathery and thick, with spiny teeth; individual leaves fall after 1 year. Flowers tiny, brownish; on long stalks; appearing with new leaves. Acorns light brown, long, conical; 1–1½″ long, one-third enclosed by thin, hairy cup; maturing in first year.

Habitat: Slopes, valleys, and oak woodlands; often in large, open groves.

Range: Coast Ranges from San Francisco to S. California and south; to 3000′.

Canyon Live Oak *Quercus chrysolepis*

Similar to the Coast Live Oak, this beautiful evergreen is slightly more northerly in distribution. It is not as tall as its cousin, but it has an enormously broad crown of stout branches that sometimes droop heavily, almost touching the ground; the crown may reach 150 feet in width. This tree is also known as the Maul Oak; its strong wood has long been used to make farm tools.

Identification — Height: 20–60′; diameter: 1–3′; sometimes larger. Massive evergreen with short trunk and large, horizontal, spreading branches; sometimes a shrub. Leaves shiny green above; paler below, sometimes with yellow hairs; elliptical, 1–3″ long and ½–1½″ wide; thick and leathery, with rounded base and short-pointed tip, and usually with spiny teeth along turned-under margins. Acorns greenish-brown; rounded at first, becoming more egg-shaped, ¾–2″ long; with shallow cup covered with yellowish hairs; maturing in second year.

Habitat — Oak woodlands, canyons, and gravelly slopes.

Range — SW. Oregon through Coast Ranges and Sierra Nevada to S. California; usually from 1000–6500′.

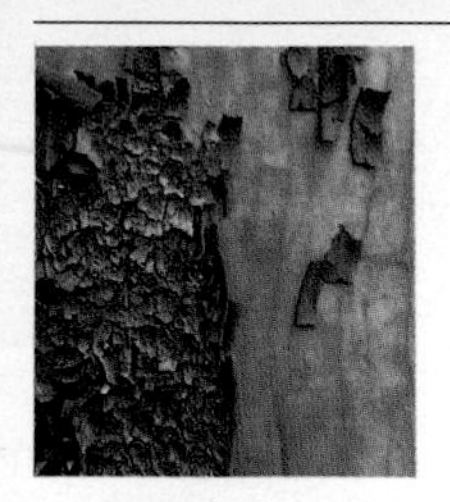

Pacific Madrone *Arbutus menziesii*

This species is the tallest American member of the heath family, which includes the rhododendrons, azaleas, and blueberries; it is also the northernmost New World heath. In spring, its long clusters of white, urn-shaped flowers, set off against the dark glossy leaves, are a cheerful sight; it also bears colorful fruit in fall.

Identification — Height: 20–80′; diameter: 2′. Tall, broad-leaved evergreen with reddish-brown trunk and stout red branches forming a narrow crown. Leaves shiny dark green above, paler below; elliptical, 2–4½″ wide and 1–3″ long; thick and leathery, sometimes with sawteeth; turning red with age. Flowers white or pinkish, urn-shaped; ¼″ long, in clusters 2–6″ long and as wide; appearing at ends of twigs in early spring. Fruit orange-red, berrylike; ⅜–½″ in diameter, with fine warts on surface; maturing in fall.

Habitat — Canyons and upland slopes, often with oaks or conifers.

Range — SW. British Columbia south along coast to S. California; also in Sierra Nevada of central California and on Santa Cruz Island; to 5000′ (sometimes slightly higher).

Curlleaf Cercocarpus *Cercocarpus ledifolius*

This small tree has dark reddish-brown heartwood, which is probably the reason that the species is also known as Curlleaf Mountain-mahogany. It is not related to the true mahogany (*Swietenia*) of the tropics, but the heavy wood is used to make souvenirs and for fuel in mining operations. In winter, elk find the foliage an important source of browse; deer eat the leaves all year.

Identification Height: 15–30′; diameter: ½–1½′. Small evergreen tree or shrub with compact aromatic crown of spreading, curved branches. Leaves shiny dark green above, paler below, and leathery; narrowly lance-shaped; ½–1¼″ long and up to ⅜″ wide; usually clustered, with curled-under edges. Flowers yellowish, hairy, and lacking petals; ⅜″ long, funnel-shaped; appearing in early spring. Fruit hairy, whitish, and seedlike; narrowly cylindrical; ¼″ long, with twisted tail 1½–3″ long.

Habitat Mountain slopes with dry, rocky soil; also in grasslands with sagebrush, pinyons, and oak, and conifer forests.

Range Extreme SE. Washington to S. Montana, south to S. California and N. Arizona; 4000–10,500′.

Velvet Ash *Fraxinus velutina*

Also called Arizona Ash, this tree is a popular ornamental in the Southwest; several horticultural varieties have been developed for use as street and shade trees. Early explorers and settlers in the Southwest knew that the presence of this tree indicated a good water source; they used the timber to build wagons.

Identification — Height: 40′; diameter: 1′. Medium-size tree with spreading branches forming an open, rounded crown. Leaves shiny green above, paler below, often with velvety hairs; opposite, pinnately compound; 3–6″ long, usually with 5 leaflets, each 1–3″ long and ⅜–1¼″ wide; leaflets lance-shaped or elliptical, with wavy margins or teeth; turning yellow in fall. Male and female flowers on different trees; ⅛″ long; in clusters; male yellowish, female greenish; appearing before leaves. Fruit a pale brown, long, narrow key, ¾–1¼″ long; in crowded clusters; maturing in summer and early fall.

Habitat — Stream banks, desert washes, canyons, and moist areas in grasslands; in oak woodlands or with Ponderosa Pine.

Range — S. Nevada to S. California and SW. Texas; 2500–7000′.

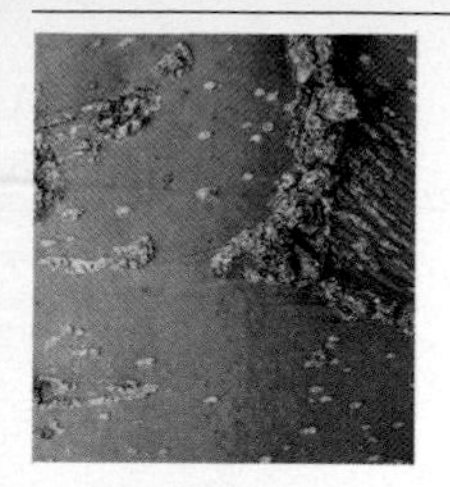

Blue Paloverde *Cercidium floridum*

The pale blue leaves of this species appear only briefly; for most of the year, the Blue Paloverde is leafless. Fortunately, its bluish twigs and branches are equipped to carry on the action of photosynthesis, using the energy of the sun to manufacture carbohydrates. The pods, seeds, and twigs are all important food sources for a variety of animals.

Identification Height: 30′; diameter: 1½′. Small, spiny tree with a bluish-green trunk and spreading crown of bluish-green branches, lacking leaves most of the year. Leaves pale blue-green; bipinnately compound, 1″ long, with 2 or 3 pairs of leaflets on each side axis; leaflets ¼″ long, oblong; shedding soon after appearing in spring. Flowers bright yellow, some with red spots; ¾″ long, with 5 petals; in 2″ clusters of 4–5; in spring, sometimes again in summer. Fruit a yellow-brown pod; oblong, flat, with 2–8 seeds; 1½–3¼″; maturing and falling in summer.

Habitat Washes, valleys, and lower desert slopes.

Range SE. California to central and S. Arizona, south to NW. Mexico; at 4000′.

Honey Mesquite *Prosopis glandulosa*

In very dry areas, the Honey Mesquite can be very shrubby, for it puts its energy into developing a large system of taproots, with which it absorbs water from the soil. The pods provide good feed for livestock, but the seeds are disseminated by the cattle that graze them; as a result, the Honey Mesquite has invaded range areas.

Identification: Height: 20′; diameter: 1′; sometimes larger. Shrub or small tree with spiny branches, short trunk, and open, spreading crown. Leaves yellow-green; bipinnately compound, 3–8″ long, with numerous pairs of stalkless leaflets; leaflets narrowly oblong, ⅜–1¼″ long and ⅛″ wide; smooth. Flowers pale yellow, fragrant; ¼″ long, in narrow, crowded clusters 2–3″ long; appearing in spring and summer. Fruit a long, narrow, light brown pod, 3½–8″ long and ¼–⅜″ wide; with several beanlike seeds; maturing in summer and remaining closed.

Habitat: Desert grasslands and deserts; in valleys, washes, and sandy plains.

Range: S. California and extreme SW. Utah east to N. Texas and NW. Oklahoma, and south to Mexico; to 4500′.

Huisache *Acacia farnesiana*

This species, also called Sweet Acacia and Cassie, is one of the few acacias native to temperate regions. In Europe, the Huisache (pronounced "we-sah′-chuh") is cultivated for its fragrant flowers, used in making perfume. A related species from Egypt, *Acacia seyal*, is reputedly the wood from which the Ark of the Covenant was made.

Identification: Height: 16′; diameter: 4″. Much-branched shrub or small tree with spiny twigs and wide crown of spreading branches. Leaves gray-green, bipinnately compound; alternate or in clusters; 2–4″ long with many oblong, stalkless, paired leaflets each ⅛–¼″ long. Flowers yellow or orange, very fragrant; 3⁄16″ long, in ball-like clusters ½″ in diameter; appearing in late winter and early spring. Fruit a cylindrical, dark brown or black pod, 1½–3″ long and ⅜–½″ thick; with many elliptical seeds; maturing in summer and remaining attached.

Habitat: Woodlands and open areas with sandy or clay soils.

Range: S. Texas; local in S. Arizona; cultivated forms across the South from Florida to California; to 5000′.

Smooth Sumac *Rhus glabra*

This shrubby species is very widespread, occurring throughout the United States in a variety of conditions, and even reaching high altitudes in the West. Smooth Sumac is also known as Scarlet Sumac, because its leaves turn bright scarlet in fall. It belongs to the cashew family, a mainly tropical group that includes cashew, pistachio, and mango trees.

Identification Height: 20′; diameter: 4″. Large shrub or small tree with sparse, spreading branches forming a small crown. Leaves shiny green above, whitish below; pinnately compound; 12″ long, with 11–31 leaflets, each 2–4″ long; leaflets lance-shaped, smooth, with sawteeth; turning red in fall. Flowers white, tiny (smaller than ⅛″), with 5 petals; in upright, crowded clusters to 8″; appearing in early summer; male and female on different plants. Fruit round, berrylike, ⅛″ in diameter; covered with sticky red hairs; in crowded clusters; maturing in late summer.

Habitat Open areas, roadsides, clearings, grasslands, and forest edges; often in sandy soil.

Range Throughout the West to 7000′.

Blue Elder *Sambucus cerulea*

The Blue Elder produces sweet berries that are used in jelly and pies. The wood is hard and smooth; watchmakers use slivers to repair and clean small, delicate mechanisms. The young twigs have thick, light pith; when this is removed, the hollow twigs can be fashioned into pop-guns and whistles.

Identification — Height: 25′; diameter: 1′. Large shrub or much-branched small tree with rounded, compact crown; often forming thickets. Leaves yellowish green above, paler below; pinnately compound, opposite; 5–7″ long, with 5–9 leaflets, each 1–5″ long and ⅜–1½″ wide, narrowly lance-shaped, with unequal base, long tip and sharp sawteeth. Flowers yellowish white, ¼″ wide, with 5-lobed corolla; fragrant, in upright clusters 4–8″ wide; appearing in summer. Fruit a dark blue berry with whitish bloom, ¼″ in diameter; in clusters; maturing in late summer or fall.

Habitat — Clearings, roadsides, forest openings, and moist stream banks and canyons.

Range — S. British Columbia to S. California along coast; in mountains from W. Montana to SW. Texas; to 10,000′.

California Buckeye *Aesculus californica*

The California Buckeye is the only member of the horsechestnut family to occur in the West. The origin of the name "buckeye" is probably rooted in the resemblance of the large seed to the eye of a deer. Like its relatives in southern Europe and Asia Minor, the California Buckeye bears sticky winter buds.

Identification Height: 25′; diameter: 1′. Small tree or shrub with short trunk and crooked branches forming a broad crown. Leaves dark green above, paler below with whitish hairs; palmately compound on long stalk; usually with 5 leaflets, each 3–6″ long and 1–2″ wide; leaflets narrowly elliptical with fine teeth; turning brown in late summer. Flowers white or pale pink, fragrant, funnel-shaped, with 4–5 petals; in upright clusters from 4–8″ long; in late spring and early summer. Fruit a pale brown, pear-shaped capsule, 2–3″ long; smooth; usually splitting into 3 parts; maturing in late summer, and containing 1 large, shiny brown, nutlike poisonous seed.

Habitat Canyons and hillsides with moist soil; oak woodlands.

Range Coast Ranges and Sierra Nevada in California; to 4000′.

Blue Oak *Quercus douglasii*

Even from far away it is easy to recognize this tree, because it has a dense crown of distinctive blue-green foliage. The wood is chiefly used as fuel, and the acorns are an important source of food for many kinds of animals. This species is also known as Mountain White Oak; it belongs to a group of oaks, known collectively as white oaks, that mature their acorns in a single year. The group called the red oaks has acorns that take two years to mature.

Identification Height: 20–60′; diameter: 1′. Small to medium-size tree with short, leaning crown and short, stout branches forming a rounded crown; sometimes a shrub. Leaves pale blue-green above, paler below and slightly hairy; oblong or elliptical, 1¼–4″ long and ¾–1¾″ wide, with 4–5 shallow lobes, with or without sawteeth. Acorns brownish, elliptical, ¾–1¼″ long; with shallow cup of bumpy scales; maturing in first year.

Habitat Dry, rocky slopes; with other oaks or with Digger Pine.

Range Restricted to California, mainly in foothills of Coast Ranges and Sierra Nevada; 300–3500′.

Oregon White Oak *Quercus garryana*

Also called Garry Oak and Oregon Oak, this species is one of the most important timber trees in the West. Its hard, durable wood is used for shipbuilding, furniture-making, cooperage, and cabinetwork, as well as for fuel. It is similar to the related White Oak (*Q. alba*), and like that eastern species is widely planted as an ornamental.

Identification — Height: 30–70′; diameter: 1–2½′. Medium-size to tall tree with stout branches forming rounded crown of dense foliage; sometimes a shrub. Leaves shiny dark green above, paler below, usually with hairs on undersurface; elliptical, rounded at both ends; 3–6″ long and 2–4″ wide, with deep, somewhat blunt lobes; sometimes turning red in fall. Acorns greenish brown to light brown, 1–1¼″ long, elliptical; one-fourth to one-third enclosed in thin cup; maturing in first year.

Habitat — Mountainsides and valleys; in pure stands or with other oaks.

Range — SW. British Columbia south in Coast Ranges and Sierra Nevada to central California; to 3000′ in north; from 1000–5000′ farther south.

California Black Oak *Quercus kelloggii*

The handsome California Black Oak can be recognized by its distinctive, deep green, bristle-tipped leaves, which are unlike the leaves of any other western oak. The edible acorns were once a staple of Native Americans from California; the wood, which is fairly brittle, is used as fuel. Also called Kellogg Oak and Black Oak.

Identification Height: 30–80′; diameter: 1–3′. Medium-size to tall tree with spreading branches forming an irregular crown. Leaves shiny deep green above, lighter yellow-green below, often with hairy underside; elliptical, 3–8″ long and 2–5″ wide; somewhat thickened; lobed, with bristle-pointed teeth; turning yellow or brown in fall. Acorn brownish, elliptical, 1–1½″ long; one- to two-thirds enclosed by scaly cup; maturing in second year.

Habitat Foothills and mountains with rocky or sandy soils; with mixed conifers or in nearly pure stands.

Range SW. Oregon to S. California in Coast Ranges and Sierra Nevada; 1000–8000′.

Bigleaf Maple *Acer macrophyllum*

One glance at the leaves will tell you how the Bigleaf Maple got its name. This species is widely planted as an ornamental and in park and garden collections; in the fall, it puts on a spectacular display of color, but the show is usually brightest and most colorful in those trees growing in their native regions. Also called Broadleaf Maple and Oregon Maple.

Identification Height: 30–70′; diameter: 1–2½′. Medium-size tree with spreading or drooping branches forming a large, rounded crown. Leaves shiny dark green above, paler and hairy below; opposite; 6–10″ long and as broad, with 5 (rarely 3) deep lobes; edges with smaller lobes and sawteeth; turning bright orange or yellow in fall. Flowers yellow, fragrant, ¼″ long, in long, drooping clusters to 6″; appearing with leaves in spring. Fruit a brownish, paired, long-winged key with stiff tan or yellow hairs; maturing in fall.

Habitat Moist soils of stream banks and canyons.

Range SW. British Columbia to S. California; to 1000′ in north; 3000–5500′ farther south.

London Planetree *Platanus × acerifolia*

In the scientific name of a plant, an "×" designates a hybrid—a cross between two or more species. The London Planetree is a cross between the Sycamore (*P. occidentalis*) and the Oriental Planetree (*P. orientalis*) of southeastern Europe and Asia. As the common name indicates, the London Planetree is a popular street tree in England; it is also found in other parts of Europe.

Identification: Height: 70′; diameter: 2′. Large tree with stout trunk and open crown of heavy, spreading branches. Leaves light green above, paler below; 5–10″ long and as wide, with 3–5 shallow, pointed lobes; sometimes with sawteeth; turning yellow or brown in fall. Flowers tiny, greenish, in spherical clusters; male and female on different twigs; appearing in spring. Fruit a rough brown ball, 1″ in diameter; in pairs from long stalk; maturing in fall, separating into many nutlets.

Habitat: Cities and suburbs in temperate regions; usually in moist soils.

Range: Throughout the West.

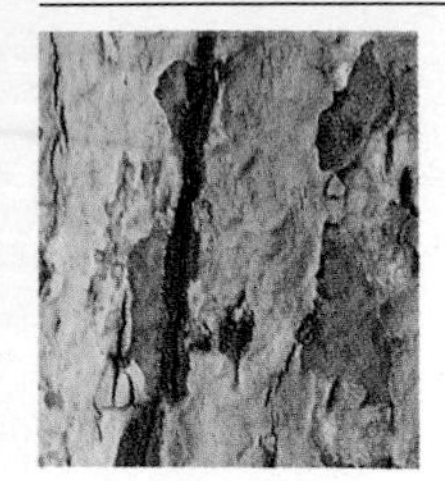

California Sycamore *Platanus racemosa*

The trunk of this massive, memorable species is often divided close to the ground; the heavy branches that result may be contorted picturesquely, or even grow along the ground for part of their length. The largest California Sycamore known grows in Santa Barbara; its trunk is more than 27 feet in circumference; the tree is 115 feet tall, with a crown almost 150 feet wide.

Identification Height: 40–80′; diameter: 2–4′; sometimes much larger. Massive tree with large trunk, often forked at base; thick, heavy branches forming broad, spreading crown. Leaves green above, paler and hairy below; star-shaped, 6–9″ long and as broad, with 3–5 deep, pointed lobes; edges wavy with a few teeth. Flowers tiny, in spherical, separate, male or female clusters; appearing in spring with leaves. Fruit a brownish ball, ⅞″ in diameter; in clusters of 2–7 on long stalks; maturing in fall and separating into nutlets in winter.

Habitat Mountains and valleys with moist soils; often along streams.

Range N. California south to Mexico; to 4000′.

Fremont Cottonwood *Populus fremontii*

The Fremont Cottonwood is common at low altitudes throughout the Southwest, especially along the Rio Grande and the Colorado River. The root of this species is used by Hopi Indians to carve Kachina dolls, powerful figures that exert an important force in the Hopi religion; the dolls are miniature likenesses of supernatural spirits, used to teach children about the Hopi pantheon.

Identification Height: 40–80′; diameter: 2–4′. Medium-size to tall tree with large, spreading branches forming an open, somewhat flattened crown. Leaves shiny yellow-green; triangular, 2–3″ long and as wide or wider; with nearly straight base and irregular, coarse teeth; turning yellow in fall. Flowers reddish, in catkins 2–3½″ long; male and female on different trees; appearing in early spring. Fruit a light brown, egg-shaped capsule, ½″ long; maturing in spring and dividing into 3 parts.

Habitat Stream banks with wet soils; often with sycamores, alders, or willows.

Range N. California east to S. and W. Colorado, south to Mexico and SW. Texas; to 6500′.

Balsam Poplar *Populus balsamifera*

Like other members of the group known as balsam poplars, this species is easy to recognize in spring, because its winter buds give off a delicious balsam fragrance just before the trees leaf out. The Balsam Poplar is widespread throughout the North; a variety known as Balm-of-Gilead is planted as an ornamental.

Identification Height: 60–80′; diameter: 1–3′. Tall tree with narrow, open crown of erect branches. Leaves shiny dark green above, whitish below, often with reddish veins; oval, with a pointed tip; 3–5″ long and 1½–3″ wide; with rounded or notched base and wavy teeth along margin. Flowers brownish, tiny, in catkins 2–3½″ long; male and female on different trees; appearing in early spring. Fruit a light brown, pointed, egg-shaped capsule, 5⁄16″ long; maturing in spring and dividing into 2 segments.

Habitat Stream banks, floodplains, and other moist areas in valleys; often in pure stands.

Range From timberline in Alaska to Newfoundland, south to SE. British Columbia, Iowa, and Pennsylvania; local in Rockies to Colorado; to 5500′.

Quaking Aspen *Populus tremuloides*

This graceful tree is emblematic of the Rocky Mountains. With the slightest encouragement from the wind, its heart-shaped leaves tremble and dance; large, pure stands of these trees create a rushing babble of summery noise. The bark is soft and smooth, equally well suited to recording the carved initials of sweethearts or the emphatic signature of a curious bear.

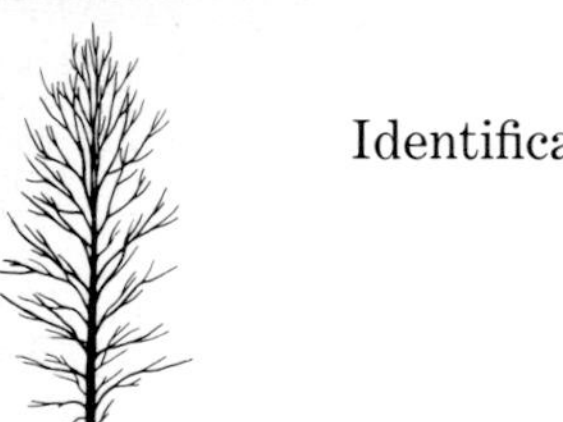

Identification Height: 40–70′; diameter: 1–1½′. Medium-size to tall tree with rounded narrow crown of rather sparse foliage. Leaves shiny green above, duller below; nearly round, with short-pointed tip; 1½–3″ long and about as wide; turning gold in fall. Flowers tiny, brownish, in catkins 1½–2″ long; male and female on different trees; in early spring before leaves. Fruit a green, conical capsule, ¼″ long, in drooping clusters to 4″ long; in late spring.

Habitat Sandy and gravelly slopes; also in moist soils; often in pure stands.

Range Alaska and Newfoundland south through mountains to S. California, S. Arizona, and Virginia; near sea level in north; 6500–10,000′ in south.

Red Alder *Alnus rubra*

Also called Oregon Alder and Western Alder, this species is the most important source of hardwood in the Northwest. The Red Alder is a pioneer on burned sites. After a forest fire, its seedlings quickly take root; they may live only briefly, but they provide the cover and stability necessary for the establishment of other trees.

Identification — Height: 40–100′; diameter: 2½′; sometimes larger. Medium-size to tall tree with pointed or rounded crown and straight trunk. Leaves dark green above, gray-green below with reddish hairs; in 3 rows along twig; oval to elliptical, somewhat rounded at ends, 3–5″ long and 1¾–3″ wide; wavy margins with double sawteeth. Flowers tiny; male yellowish, in cylindrical, drooping catkins 4–6″ long; female reddish, in slender cones ⅜–½″; before leaves. Cones blackish, ½–1″; in short-stalked clusters of 4–8; mature late in summer and remain attached.

Habitat — Stream banks and lower slopes with moist soils; often in pure stands, sometimes forming thickets.

Range — SE. Alaska along coast to central California; local in N. Idaho; to 2500′.

Water Birch *Betula occidentalis*

Although it is comparatively rare, the Water Birch has a wide distribution in the West. It tends to be shrubby, especially where it occurs as an ornamental. Other western relatives of this tree have a more northerly range; this species is the only birch that occurs in the Southwest. Also known as Red Birch and Black Birch.

Identification Height: 25′; diameter: 6–12″. Small tree or shrub with spreading, drooping branches. Leaves dark green above, paler yellowish green below with tiny gland-dots; oval, 1–2″ long and ¾–1″ wide, with double sawteeth along margin; turning yellowish tan in fall. Flowers tiny; male flowers yellowish in long catkins; female flowers in shorter, erect catkins on same twig. Cones brownish, cylindrical, 1–1¼″ long; growing erect on stalk; maturing in late summer.

Habitat Mountain canyons with conifers, cottonwoods, and willows; usually along stream banks.

Range NE. British Columbia east to S. Manitoba, south to California and N. New Mexico; 2000–8000′.

Black Cottonwood *Populus trichocarpa*

Also known as the Western Balsam Poplar, this species is the largest deciduous tree of the Northwest and the tallest cottonwood. In early spring, the buds and young leaves exude a strong balsam fragrance. The wood is used to make crates, pulp, and veneer. The Black Cottonwood is a common street tree in England.

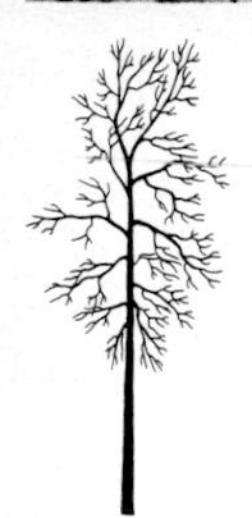

Identification — Height: 60–120′; diameter: 1–3′; sometimes much larger. Tall tree with upright branches forming a narrow, open crown. Leaves shiny dark green above, whitish below, often with reddish veins; broadly oval, 3–6″ long and 2–4″ wide; with pointed tip and rounded base; margins with fine, wavy teeth; turning yellow in fall. Flowers tiny, purplish red; in catkins 1½–3¼″ long; male and female on different trees; appearing in early spring. Fruit a light brown, rounded capsule, ¼″ in diameter; hairy, maturing in spring and dividing into 3 parts.

Habitat — Stream banks, floodplains, and other moist areas; in pure stands or with willows and Red Alder.

Range — S. Alaska to S. California; in Rockies to extreme SW. Alberta and N. Utah; to 2000′ in north, to 9000′ in south.

Common Chokecherry *Prunus virginiana*

Widespread throughout much of North America, the Common Chokecherry produces bitter fruit. These chokecherries are edible despite their flavor, and some people use them to make jelly or jam—but great caution is advised, as the stones are poisonous. Also called Western Chokecherry and (in the East) Eastern Chokecherry, this species has three rather indistinct geographical varieties.

Identification: Height: 20′; diameter: 6″. Small tree with open crown or thicket-forming shrub. Leaves shiny dark green above, paler and occasionally hairy below; elliptical, 1½–3¼″ long and ⅝–1½″ wide; margins with sharp sawteeth; turning yellow in fall. Flowers white, ½″ wide, with 5 rounded petals; appearing in clusters, to 4″ wide, in late spring. Fruit dark red or blackish, cherrylike, with large, poisonous stone; maturing in summer.

Habitat: Streamsides, clearings, forest edges, and mountains with moist soils.

Range: N. British Columbia to S. California, east to North Carolina and Newfoundland; to 8000′ in the Southwest.

Blueblossom *Ceanothus thyrsiflorus*

The emblem of Bacchus, the Greek god of wine and revelry, was a *thyrsus*—a staff adorned with flower clusters and ivylike leaves. The scientific name of this species recalls that emblem and refers to the dense, branched flower clusters borne in spring by the Blueblossom. In spring, these colorful flowers are a familiar sight along the highways of California.

Identification — Height: 20′; diameter: 8″. Evergreen shrub or small tree with many spreading branches arising from short trunk. Leaves shiny green above, paler below with some hairs; elliptical to oblong, ¾–2″ long and ½–¾″ wide; somewhat rounded at ends, with fine, wavy sawteeth. Flowers fragrant, light to deep sky blue, 3⁄16″ wide with 5 petals; in crowded clusters 1–3″ long; appearing in spring. Fruit a sticky black capsule, 3⁄16″ in diameter, smooth, 3-lobed and dividing into 3 parts; maturing in summer.

Habitat — Mixed conifer forests, chaparral, and Redwood groves in canyons and mountainsides.

Range — Along the Pacific Coast from SW. Oregon to S. California; to 2000′.

Pacific Dogwood *Cornus nuttallii*

Like many other dogwoods, this species is especially beautiful in spring, when its showy pink and white flowers light up the landscape. The name "dogwood" originated in Europe, where the Common Dogwood (*C. sanguinea*) was used to make "dogs"—wooden spikes or skewers; the term survives today among loggers.

Identification — Height: 50′; diameter: 1′; rarely somewhat larger. Medium-size tree with dense crown of horizontal branches. Leaves shiny green and smooth above, paler and hairy below; elliptical, 2½–4½″ long and 1¼–2¾″ wide; with distinct veins and slightly wavy edges; turning orange or red in fall. Blossoms white or pink, 4–6″ wide, with 6–7 petal-like bracts; actual flowers ¼″ long, greenish, in crowded heads at center of bracts; appearing in late spring or early summer, sometimes again in fall. Fruit red or orange, berrylike, ½″ wide; in crowded clusters to 1½″ wide; maturing in fall.

Habitat — Conifer forests with moist soils; generally in mountains.

Range — SW. British Columbia south in mountains to S. California; local in Idaho; to 6000′.

Tanoak *Lithocarpus densiflorus*

The Tanoak looks like something of a botanical mongrel: Its flowers are like a chestnut's, but it bears acorns like an oak, and in fact, it is sometimes called the California Chestnut-oak. The bark of this tree is an important source of tannin, which is used to tan leather; this fact accounts for yet another common name, Tanbark-oak.

Identification: Height: 50–80′; diameter: 1–2½′. Medium-size to tall evergreen with broad, conical, or irregular crown. Leaves covered with yellowish wool when young, becoming shiny light green above, with whitish or yellow hairs below; oblong, 2½–5″ long and ¾–2¼″ wide, with wavy edges; thick and leathery. Flowers tiny, whitish, in catkins 2–4″ long; mostly male; a few tiny female flowers present below catkins; appearing in early spring and early fall. Acorns light brown, egg-shaped, ¾–1¼″ long; 1–2 on stalk; with shallow cup covered with long scales; maturing in second year.

Habitat: Moist valley soils; with oaks or in almost pure stands.

Range: SW. Oregon to S. California along coast; in Sierra Nevada to central California; to 5000′.

Toyon *Heteromeles arbutifolia*

This beautiful, showy plant is also known as Christmas-berry and California-holly, for it bears plentiful red fruit and is a popular holiday decoration. The Toyon, which belongs to the rose family, is the only member of the genus *Heteromeles;* it was formerly placed with the genus *Photinia* in the same family. There are several varieties in cultivation.

Identification Height: 30′; diameter: 1′. Small evergreen tree or shrub with short trunk and many branches forming a rounded crown. Leaves shiny dark green above, paler below; thick; oblong to elliptical, 2–4″ long and ¾–1½″ wide, with sharp sawteeth. Flowers white, ¼″ wide, with 5 petals; in crowded, upright clusters 4–6″ wide; appearing in early summer. Fruit like a tiny red or yellow apple, ¼–⅜″ in diameter; maturing in fall and persisting in winter.

Habitat Chaparral and woodland areas, along streams and on dry hillsides; sometimes on coastal cliffs.

Range Coast Ranges and foothills of Sierra Nevada in California; also Channel Islands; to 4000′.

Scouler Willow *Salix scoulerana*

The Scouler Willow is quick to establish itself on burned-over areas, giving rise to the alternate name, Fire Willow. It is one of the earliest willows to flower in spring, and is browsed by moose and livestock. This species is able to tolerate remarkable extremes of hot and cold, from winter temperatures of −75° F to summer heat of 120° F.

Identification — Height: 15–50′; diameter: 1½′. Small tree or shrub with rounded, compact crown and straight trunk. Leaves dark green and smooth above, whitish with gray or reddish hairs below; usually elliptical (but variable), 2–5″ long and ½–1½″ wide, usually widest beyond middle and with short tip; edges smooth or wavy-toothed. Flowers in catkins 1–2″ long, covered with hairy blackish scales; stout; in early spring before leaves. Fruit a light brown, hairy capsule; ⅜″ long; maturing in early summer.

Habitat — Clearings, cut- or burned-over areas, and forest openings in understory of upland coniferous forests.

Range — Central Alaska to Manitoba, south to California and New Mexico; also in Black Hills; to 10,000′ in mountains.

Bebb Willow *Salix bebbiana*

The Bebb Willow, named for the American botanist Michael Shuck Bebb, is widespread throughout much of the north country. It is an important member of a group known as "diamond willows," which develop diamond-shaped patterns on their trunks in response to a fungal infection. According to some experts, the inner bark is a fine material for making fishing lines.

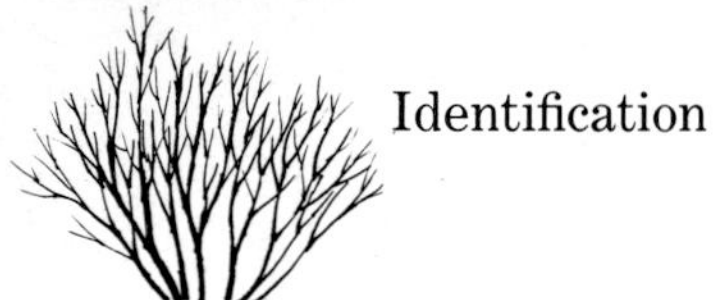

Identification: Height: 10–25′; diameter: 6″. Many-branched small tree or shrub with broad, rounded crown. Leaves dull green above, whitish or gray below, with pattern of veins; elliptical with short-pointed ends; 1–3½″ long and ⅜–1″ wide; with wavy or slightly toothed edges. Flowers yellow or brown, in scaly catkins ¾–1½″ long; appearing on short stalks in spring, sometimes before leaves. Fruit a slender, hairy, light brown capsule, ⅜″ long, with long stalk and long point at tip; maturing in early summer.

Habitat: Along streams, lakes, and swamps; also in moist open upland areas.

Range: Central Alaska south to British Columbia; in Rockies to S. New Mexico, east to Newfoundland and Maryland.

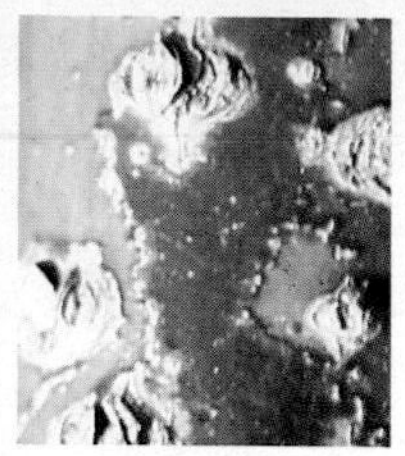

Red-osier Dogwood *Cornus stolonifera*

Often planted as an ornamental, the Red-osier Dogwood is one of the most widely distributed shrubs in all of North America. It is handsome at all times of year: It flowers in spring, bears showy fruit in fall, and in the winter it enlivens the landscape with its unusual reddish twigs. Also called Red Dogwood and Kinnikinnik.

Identification: Height: 3–10′; diameter: 3″; rarely somewhat larger. Large, thicket-forming shrub with many stems; rarely a small tree. Leaves dull green above, whitish below with fine hairs; oval or elliptical, pointed, with smooth edges and distinctive parallel, curved veins; turning red in fall. Flowers white, ¼″ wide, with 4 petals; in crowded clusters to 2″ wide; appearing in late spring and early summer. Fruit berrylike, whitish; maturing in late summer.

Habitat: Stream banks and other areas with moist soils; often in mixed forests.

Range: Central Alaska to California, east to Newfoundland and Virginia; to 5000′; to 9000′ in the Southwest.

Pacific Willow *Salix lasiandra*

This widespread willow is found in the Sierran montane forest as well as in mountains along the Pacific Coast. Also known as Western Black Willow and Yellow Willow, the species is a familiar sight along riverbanks. The closely related Bonpland Willow (*S. bonplandia*) is similar in appearance; it is found in wet soils in California, Utah, and southern Arizona.

Identification: Height: 20–50′; diameter: 2′. Small tree or thicket-forming shrub with irregular, open crown. Leaves shiny green above, whitish below; narrowly lance-shaped, 2–5″ long and ½–1″ wide, with rounded base and long-pointed tip; edges with fine sawteeth. Flowers in yellow or brown catkins, 1½–4″ long, with hairy scales; appearing with leaves in spring. Fruit a light reddish-brown capsule, ¼″ long; smooth; maturing in early summer.

Habitat: In wet soils along lakes and streams; also along roads; on mountain slopes and valleys.

Range: SE. Alaska and Saskatchewan south in mountains to S. California and S. New Mexico; to 8000′.

Oregon Ash *Fraxinus latifolia*

This species produces valuable wood that is used to make furniture, floors, and boxes, as well as in millwork and paneling. According to legend, venomous snakes find the Oregon Ash uncongenial, and a branch or a stick will ward the creatures off. This species is the only ash that is native to the northwestern coast.

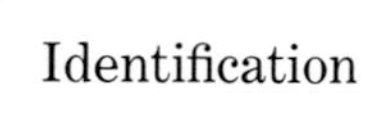

Identification — Height: 80′; diameter: 2′. Tall tree with straight trunk and narrow, dense crown. Leaves light green above, paler and hairy below; opposite and pinnately compound, 5–12″ long, with 5–7 leaflets; each elliptical, 2–5″ long and 1½″ wide; short-pointed at both ends; sometimes with a few sawteeth. Flowers ⅛″ long; male yellowish, female greenish, in clusters on separate trees; appearing before leaves in early spring. Fruit a light brown key, 1¼–2″ long, with rounded wing; in dense, crowded clusters; maturing in late fall.

Habitat — Stream banks and canyons with wet soils; often with willows, Red Alder, or Black Cottonwood.

Range — W. Washington south along coast to N. California, and in Sierra Nevada to central California; to 5500′.

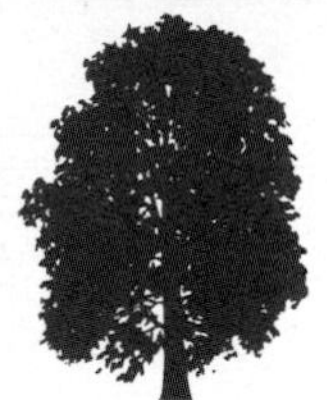

California-laurel *Umbellularia californica*

This green-barked member of the laurel family is a popular ornamental and shade tree. The light brown wood with dark streaks is prized for veneers, furniture, and cabinetry. The leaves and twigs have a pungent, peppery aroma, giving rise to the alternate name Pepperwood; because the smell is said to bring on headaches, the species is also called the Headache Tree.

Identification: Height: 40–80′; diameter: 1½–2½′. Medium-size to tall evergreen (sometimes a shrub) with short, forking trunk and large, dense, rounded crown of spreading branches. Leaves shiny dark green above, paler below; leathery and thick; elliptical, 2–5″ long and ½–1½″ wide, with short-pointed or rounded tip; fading to yellowish and shedding after second year. Flowers pale yellow, ¼″ long, in rounded clusters of about 10; in late winter or early spring. Fruit a green to purple berry, ¾–1″ long, with thin pulp and large seed; maturing in late fall.

Habitat: Canyons and valleys with moist soils; in mixed forests.

Range: SW. Oregon to S. California in Coast Ranges and Sierra Nevada; to 4000′ in most places, 2000–6000′ in south.

Bitter Cherry *Prunus emarginata*

Songbirds and small mammals appear not to notice the intensely bitter flavor of this tree's fruit, for they consume the cherries in great quantities. The leaves, also bitter, form important browse for deer and livestock. The most common cherry of the West, it is also known as Quinine Cherry and Wild Cherry.

Identification Height: 20′; diameter: 8″. Small tree or thicket-forming shrub with upright branches and rounded crown. Leaves dark green above, paler below; oblong to elliptical, 1–2½″ long and ⅜–1¼″ wide; with round or blunt tip and short-pointed base with 1–2 gland-dots; edges with fine, blunt, gland-tipped sawteeth. Flowers white, ½″ wide, with 5 rounded and notched petals; in groups of 3–10 on thin stalks; in spring with leaves. Fruit a red to blackish cherry, 5⁄16–⅜″; round, bitter; maturing in summer.

Habitat Valleys and mountainsides with moist soils; also in coniferous forests and chaparral.

Range W. British Columbia south to S. California; also W. Montana, E. Oregon to E. California and south to SW. New Mexico; to 9000′.

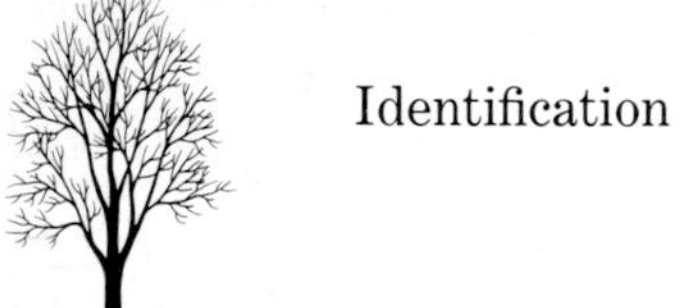

Narrowleaf Cottonwood *Populus angustifolia*

Closely related to the Balsam Poplar (*P. balsamifera*), but with willowlike leaves, the Narrowleaf is the most common cottonwood of the northern Rockies. Also known as Mountain Cottonwood and Black Cottonwood, it was first discovered by Lewis and Clark in the course of their expedition. In spring, the buds are coated with resin and give off a delightful balsam fragrance.

Identification: Height: 50′; diameter: 1½′. Tall slender tree with narrow conical crown of upright branches. Leaves shiny green above, paler below; lance-shaped with long-pointed tip and rounded base, 2–5″ long and ½–1″ wide; edges with fine sawteeth; turning dull yellow in fall. Flowers reddish, in catkins 1½–3″ long; male and female on different trees; appearing in early spring before leaves. Fruit a light brown, egg-shaped capsule, about ¼″ long; smooth; dividing into 2 parts; maturing in spring.

Habitat: Along mountain streams, often with willows and alders in coniferous forests.

Range: S. Alberta and adjacent Saskatchewan, south in mountains to California and SW. Texas; 3000–8000′.

Bluegum Eucalyptus *Eucalyptus globulus*

Introduced from Tasmania, the Bluegum Eucalyptus is one of the most extensively planted ornamentals in the world. It has adapted to life in Europe, China, and many parts of South America; in Ethiopia, this tree has been in large part responsible for the stabilization of nomadic populations, providing an invaluable source of fuel in an unforgiving countryside. In North America, it is seen mostly in California.

Identification: Height: 120′; diameter: 3′. Tall, straight-trunked evergreen with narrow, irregular crown; foliage has odor of camphor. Leaves dull blue-green to green; narrowly lance-shaped, 4–12″ long and 1–2″ wide, with long point at tip; usually curved; leathery and thick; leaves on young trees bluer with whitish bloom. Flower a spherical cluster of whitish stamens, without petals; scattered singly and appearing in winter and spring. Fruit a bluish-white capsule, top-shaped, ¾–1″ long; upper surface warty with 3–5 openings; in winter and spring.

Habitat: Subtropical regions with moist soils.

Range: Widely planted in California; becoming naturalized.

Desert-willow *Chilopsis linearis*

Despite its common name, this species is not related to the true willows (*Salix*); it belongs to the bignonia family, a large group of mostly tropical woody vines, shrubs, and trees, and is the only member of the genus *Chilopsis*. The wood is used for fence posts and fuel; the flexible, sturdy twigs are well suited for baskets.

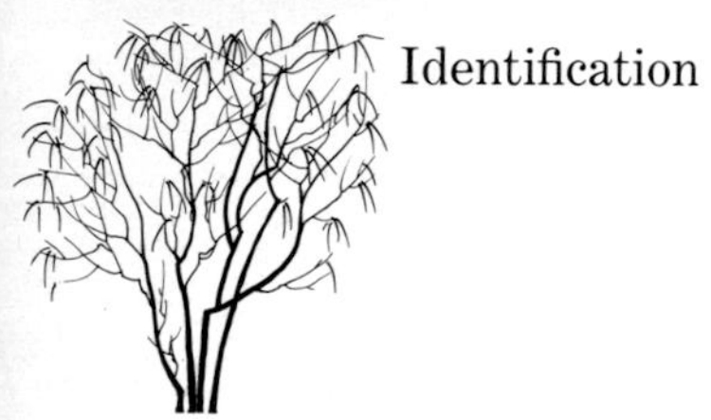

Identification — Height: 25′; diameter: 6′. Small tree or large shrub with open, spreading crown and often with leaning trunk. Leaves light green and willowlike; 3–6″ long and ¼–⅜″ wide; linear or slightly curved, drooping, sometimes hairy or sticky. Flowers whitish with pink or lavender tints; bell-shaped, 1¼″ long, with 5 unequal lobes; fragrant; appearing in clusters up to 4″ long in late spring or early summer. Fruit a dark brown capsule; cigar-shaped, 4–8″ long and ¼″ thick; maturing in fall and dividing into 2 parts; remaining attached in winter.

Habitat — Stream banks and other moist areas in plains, foothills, and desert grasslands; often forming thickets.

Range — S. California and SW. Utah to New Mexico and SW. Texas; 1000–5000′.

California Washingtonia *Washingtonia filifera*

Today, this palm is found only in a few isolated groves, but scientists believe that there were vast forests of this tree in North America many thousands of years ago. The California Washingtonia (named for George Washington) is the tallest palm native to the United States. It is easily recognized by the drooping dead leaves that clothe the trunk. Also known as California-palm and Fanpalm.

Identification: Height: 20–60′; diameter: 2–3′. Tall evergreen palm with massive trunk and large, fan-shaped leaves. Leaves gray-green, 3–5′ long; bladelike, with many leathery segments and frayed edges. Flowers white, funnel-shaped, ⅜″ long; slightly fragrant; in clusters 6–12″ long; appearing late in spring. Fruit a black berry, ⅜″ in diameter; edible; maturing in early fall.

Habitat: Colorado and Mojave deserts, in alkaline soils along streams.

Range: Restricted to small areas of SE. California and SW. Arizona; 500–3000′.

Soaptree Yucca *Yucca elata*

Although it looks like a palm, the Soaptree Yucca is actually a member of the lily family. The common name of this species refers to a material in the trunk and roots, which has been used as a substitute for soap. Native Americans of the Southwest used to cook and eat the flowers and buds of this yucca, which is found in the Sonoran and Chihuahuan deserts.

Identification: Height: 10–17′; diameter: 6–10″; sometimes larger. Palmlike evergreen small tree or shrub with long leaves and sometimes with several upright branches. Leaves yellowish green with white margins; leathery; long and narrow, 1–2½′ long and ⅛–⅜″ wide; with sharp spine at tip. Flowers white, bell-shaped; 1½–2″ long, with 6 broad sepals; in crowded clusters on long stalk; clusters 3–10′ tall; appearing in spring. Fruit a light brown, cylindrical capsule, 1½–3″ long; dry; maturing in early summer; opening into 3 parts and remaining attached.

Habitat: Deserts and desert grasslands, sandy plains, and washes.

Range: Central Arizona to SW. Texas; local in SW. Utah; 1500–6000′.

Joshua-tree *Yucca brevifolia*

A common species of the Mojave Desert, the Joshua-tree becomes contorted and misshapen with age. To the early Mormon settlers, the outline suggested a supplicant in the wilderness, and they called it the Joshua-tree in reference to Moses' successor, who led the Israelites to Canaan. Like other yuccas, the Joshua-tree relies on the nighttime activities of yucca moths for pollination.

Identification: Height: 15–30′; diameter: 1–3′. Small evergreen with short trunk and open crown of thick, upright branches. Leaves blue-green with yellowish margins; sword-shaped, 8–14″ long and ¼–½″ wide; stiff, flat, with keel on outer surface; edges with tiny teeth; sharp spine at tip. Flowers greenish yellow, bell-shaped; 1¼–2½″ long, with 6 sepals; in crowded, upright clusters 1–1½′ long; with pungent odor; in early spring. Fruit a green or brown capsule; elliptical, 2½–4″ long, 2″ thick; maturing in late spring and shedding, but remaining undivided.

Habitat: Dry desert soils, plains, and mesas; often in groves.

Range: Mojave Desert: extreme SW. Utah, Nevada, California, and Arizona; 2000–6000′.

Jumping Cholla *Opuntia fulgida*

Like other cactuses, the Jumping Cholla has sharp, barbed spines that can cause painful wounds. The spine-bearing joints are easily detached from the plant, giving unwary desert visitors the impression that they jump out at them. *Cholla* (pronounced "chó ya") is Spanish for "head"; it is a common name given to several desert cactuses.

Identification: Height: 15′; diameter: 6″. Shrubby cactus (sometimes a small tree), usually leafless, with stout, spreading branches. Leaves (present only on young plants) light green, cylindrical; ½–1″ long; fleshy, long-pointed. Flowers pink or white, streaked with lavender, 1″ long and as wide, with 5–8 petals; appearing on joints or on fruits in late spring and summer. Fruit green, pear-shaped, 1–1⅜″ long and ¾″ in diameter; fleshy; hanging in long, branched clusters; maturing in spring and summer, sometimes persisting for several years.

Habitat: Deserts, dry plains, and sandy valleys and slopes; sometimes in dense groves.

Range: Central and S. Arizona south to Mexico; to 4000′.

Saguaro *Cereus giganteus*

Rising tall and green against the cloudless blue of the desert sky, the Saguaro is the largest cactus in North America. In the desert, where snakes, rodents, and other predators are legion, safe nesting space is at a premium. Desert woodpeckers find it easy to drill a hole in the succulent branches of the Saguaro and carve out their nests; in subsequent seasons, these cavities are appropriated by other birds.

Identification: Height: 20–35′; diameter: 1–2′; sometimes larger. Tall, erect, massive cactus with several upright, thick branches. Leafless; trunk and branches yellow-green, ridged, with clusters of gray-green spines. Flowers white, funnel-shaped, 4–4½″ long and 2–3″ wide; with many waxy petals; appearing in late spring and sometimes in late summer near tops of branches. Fruit a red, egg-shaped berry, 2–3½″ long; maturing in early summer; opening to resemble flowers.

Habitat: Desert foothills with rocky soils; often on south-facing slopes; frequently with paloverdes.

Range: S. Arizona to Mexico; local in SE. California; 700–3500′.

Guide to Families

"What tree is that?" To a beginner, learning to distinguish among the hundreds of tree species in North America may seem a formidable task. All species of trees belong to a genus and a family, however, and learning to recognize the broad, shared features of these groups is a good shortcut to identification.

Pines

All members of the pine family have needlelike leaves and woody cones that contain the seeds. Within the family, each genus displays additional features. Pines (*Pinus*) have their needles clustered in groups of two to five (except for one species of pinyon pine, which has single needles). The soft, or white, pines have clusters of five needles, while the hard, or yellow, pines have clusters of two or three needles and cones with prickles or spines on the scales. Spruces (*Picea*) have individual, sharply pointed needles borne on tiny wooden pegs. Hemlocks (*Tsuga*) have small needles borne in dense sprays and small, pendent cones. True firs (*Abies*) have upright cones borne high in the tree crown; these disintegrate at maturity, shedding both seeds and cone-scales and leaving a central spike. Larches (*Larix*), which are deciduous, bear needles in pincushionlike clusters on short shoots. Douglas-firs (*Pseudotsuga*) have sharply pointed, reddish-brown buds and pendent cones with pitchfork- or trident-shaped bracts.

Cypresses A second large family of conifers is the cypress, or "cedar," family. (The true cedars are in the genus *Cedrus*, and are actually members of the pine family.) The redcedars or arborvitae (*Thuja*), white-cedars or false-cypress (*Chamaecyparis*), incense-cedars (*Libocedrus*), and cypresses (*Cupressus*) are genera with scalelike foliage so closely attached as to cover the twig. The fruits are small, woody cones of varying structure and appearance. The junipers (*Juniperus*), widespread in arid woodlands, have awl-like leaves and berrylike cones.

Redwoods The redwood family has two distinct members in the West—Coast Redwood and Giant Sequoia, which have different foliage (needlelike in the redwood and awl-like in the sequoia), but similar egg-shaped woody cones. Both species can grow to huge size and venerable age. The yew family also has needlelike leaves; its seeds have a fleshy or fruitlike outer covering.

Hardwood Families Although conifers are predominant in the West, many important broadleaf families are represented, in some instances by evergreen species.

Beeches All members of the beech family have simple, alternate leaves and a nut wholly or partially enclosed within some kind of husk. The most important genus is oak (*Quercus*); all species have a husk that partially encloses the nut and is recognized as the cap of the acorn. Oak leaves may be deciduous and lobed (the classic "oak leaf") or leathery and evergreen, lacking lobes and sometimes with spiny margins (the live oaks). Chinkapins (*Castanea*) have leathery, willow-shaped evergreen leaves and a round, very prickly or spiny fruiting structure that contains several nuts. Tanoak (*Lithocarpus*), widespread in Asia, is represented here by a single acorn-bearing species.

Willows The willow or poplar family is an important collection of fast-growing species typically found along streams and rivers. All are deciduous and bear their flowers in catkins (pussy willows are catkins), but the leaves of willows (*Salix*) tend to be long and narrow, while those of poplars and aspens (*Populus*) are typically triangular.

Birches Simple, deciduous leaves and catkins are common features of alders (*Alnus*), hophornbeams (*Ostrya*), and birches (*Betula*), all members of the birch family. Older alders often develop whitish bark, especially along the Pacific Coast, and all of them favor moist habitats.

Maples The maples (*Acer*) are easily recognized by the familiar palmately lobed leaf (the symbol of Canada) and the distinctive winged seed, borne in pairs joined at the base. Boxelder has pinnately compound leaves rather than the classic maple leaf, but still bears winged fruit.

Laurels and Sycamores The laurel family is represented by California-laurel (*Umbellularia*). This fragrant, evergreen hardwood generally occurs on wet sites, such as along streams. The sycamore family includes only a single genus, *Platanus*. Sycamores are easy to recognize, especially older trees: The outer bark peels away, revealing patches of the smooth, light inner bark.

Roses The rose family includes many genera important as crop plants and ornamentals—cherries and plums (*Prunus*), apple (*Malus*), pear (*Pyrus*), mountain-ash (*Sorbus*), serviceberry (*Amelanchier*), and hawthorn (*Crataegus*), among others. Most members of this diverse family have pretty flowers and edible fruits. The apples and their allies bear a pome—a large fleshy fruit with seeds in a papery chamber; the cherries and their allies have a fleshy, one-seeded fruit that does not split open. Many important shrubs of arid lands, such as mountain-mahogany (*Cercocarpus*), also belong to this family.

Legumes and Cashews Worldwide, the legume or pea family includes a very large number of shrub and tree species. Many members are easily recognized by their pinnately compound leaves and the fruits, which are legumes (a papery version of a pea pod with hard, dark seeds). Mesquites (*Prosopis*), paloverdes (*Cercidium*), and acacias (*Acacia*) all belong to this family. The cashew family is mostly tropical, but the sumacs (*Rhus*) are common in temperate regions. They are distinguished by their pinnately compound leaves, which are often very bright in the fall. A closely related genus, *Toxicodendron*, includes toxic species such as poison oak, poison ivy, and poison sumac.

Buckeyes and Dogwoods Represented by a single native species in the West, the buckeye or horsechestnut family is known for the large, palmately compound leaves and large nuts enclosed in a leathery capsule. Dogwoods are a variable family of shrubs and small trees, best known for those species (*Cornus*) with showy flowers. Most dogwoods have a one-seeded fleshy fruit.

Heaths and Olives A rich variety of shrubs and trees, many with attractive flowers, smooth reddish bark, and glossy, evergreen leaves, makes up the heath family. Included are rhododendrons and azaleas (*Rhododendron*), manzanitas

(*Arctostaphylos*), and madrones (*Arbutus*). Pacific Madrone, with its smooth, reddish inner bark and shiny evergreen leaves, is the most conspicuous western tree in this family. The olive family is represented by the ashes (*Fraxinus*), with opposite, pinnately compound leaves and seeds with an elongated terminal wing.

Desert Trees

Most of us can recognize members of the cactus, lily, and palm families. Cacti in the genera *Opuntia* and *Cereus* have fleshy, succulent stems that store water and are the sites for photosynthesis; the leaves in most species are reduced to spines. Palms (*Washingtonia*) take a variety of forms but all have sprays of palm fronds at the top of the stem. The lilies are mostly small herbs, but include some shrubs and small trees—notably the yuccas, with unbranched trunks and long, narrow, pointed leaves.

As you see, it is easy to place many trees species into categories of families and genera with just a few characteristics. By this process you can find a general answer to the question, "What tree is that?" Don't be discouraged by exceptions to any rule of identification!

Glossary

Alternate
Single along a twig or shoot; not whorled or in pairs.

Catkin
A compact cluster of reduced, stalkless, and usually unisexual flowers.

Cone-scale
One of the overlapping, seed-bearing scales of a cone.

Crown
The branches, twigs, and leaves at the top of a tree.

Deciduous
Shedding leaves seasonally, leafless for part of the year.

Drupe
A fleshy fruit with a central, stonelike core containing 1 or more seeds.

Genus
A group of closely related species. Plural, genera.

Introduced
Established in an area by man; exotic or foreign.

Opposite
In pairs along a twig or shoot, with 1 on each side; not alternate or whorled.

Persistent
Remaining attached, and not falling off.

Sheath
In some conifers, the papery tube enclosing the bases of needles.

Shrub
A woody plant, smaller than a tree and with several stems arising from a single base.

Species
A group of plants or animals composed of individuals that interbreed and produce similar offspring.

Tree line
The upper limit of tree growth at high latitudes or on mountains; timberline.

Wing
A thin, flat, dry projection on a fruit or seed.

ndex

hotographers
ll photographs were taken by
avid Cavagnaro with the
xception of those listed below.

onja Bullaty and Angelo Lomeo
27, 34, 35, 45, 46, 47, 48, 49, 52,
3, 55, 56, 64, 77, 83, 86, 88, 89,
10, 111, 112, 124, 125, 130, 131,
36, 137, 140, 141, 154, 155)
cooter Cheatham (107, 108, 109)
amela J. Harper (70)
Villiam Jordan/University of
Visconsin Arboretum (152, 153)
etty Randall and Robert Potts
25, 29, 30, 31, 33, 42, 43, 51, 57,
8, 59, 60, 61, 62, 63, 65, 66, 67,
9, 97, 100, 101, 103, 128, 129, 132,
39, 157, 164, 165, 176)
lark Schaack (173, 177)
ohn J. Smith (179)
ichard Spellenberg (23, 37, 68,
6, 99, 170, 172, 174)
teven C. Wilson/Entheos (178)

over Photograph
ort-Orford-cedar by
etty Randall and Robert Potts

Illustrators
All of the tree silhouettes were
drawn by Dolores R. Santoliquido
except for those listed below.

Stephen Whitney (22, 28, 36, 40, 44, 54, 58, 72, 76, 78, 86, 134, 136, 150, 152)
Daniel Allen (70, 146)
All other line drawings by Bobbi Angell.

All editorial inquiries should be addressed to:
Chanticleer Press
568 Broadway, Suite #1005A
New York, NY 10012
(212) 941-1522

To purchase this book, or other National Audubon Society illustrated nature books, please contact:
Alfred A. Knopf, Inc.
201 East 50th Street
New York, NY 10022
(800) 733-3000

Prepared and produced by
Chanticleer Press, Inc.
Publisher: Andrew Stewart
Managing Editor: Edie Locke
Art Director: Amanda Wilson
Production Manager: Susan Schoenfeld
Photo Editor: Giema Tsakuginow
Photo Assistant: C. Tiffany Lee
Publishing Assistant: Alicia Mills

Founding Publisher: Paul Steiner

Staff for this book:

Editor in Chief: Gudrun Buettner
Executive Editor: Susan Costello
Managing Editor: Jane Opper
Senior Editor: Ann Whitman
Natural Science Editor: John Farrand, Jr.
Associate Editor: David Allen
Assistant Editor: Leslie Marchal
Production: Helga Lose, Gina Stead
Art Director: Carol Nehring
Art Associate: Ayn Svoboda
Picture Library: Edward Douglas

Original series design by
Massimo Vignelli

NATIONAL AUDUBON SOCIETY

The mission of the NATIONAL AUDUBON SOCIETY *is to conserve and restore natural ecosystems, focusing on birds and other wildlife for the benefit of humanity and the Earth's biological diversity.*

We have 500,000 members and an extensive chapter network, plus a staff of scientists, lobbyists, lawyers, policy analysts, and educators. Through our sanctuaries we manage 150,000 acres of critical habitat.

Our award-winning *Audubon* magazine, sent to all members, carries outstanding articles and color photography on wildlife, nature, and the environment. We also publish *Audubon Field Notes,* a journal reporting seasonal bird sightings continent-wide; *Audubon Activist,* a newsjournal; and *Audubon Adventures,* a newsletter reaching 600,000 elementary school students. Our *World of Audubon* television shows air on TBS and public television.

For information about how you can become a member, please write or call the Membership Department at:

NATIONAL AUDUBON SOCIETY
700 Broadway
New York, New York 10003
(212) 979-3000